# Educating Scientists and Engineers for Academic and Non-Academic Career Success

# Educating Scientists and Engineers for Academic and Non-Academic Career Success

## JAMES G. SPEIGHT

CRC Press
Taylor & Francis Group
Boca Raton London New York

CRC Press is an imprint of the
Taylor & Francis Group, an **informa** business

CRC Press
Taylor & Francis Group
6000 Broken Sound Parkway NW, Suite 300
Boca Raton, FL 33487-2742

Printed on acid-free paper
Version Date: 20141007

International Standard Book Number-13: 978-1-4665-5356-9 (Paperback)

**Visit the Taylor & Francis Web site at**
**http://www.taylorandfrancis.com**

**and the CRC Press Web site at**
**http://www.crcpress.com**

# Contents

# Preface

The education of scientists and engineers is an activity of increasingly growing importance in the modern technologically oriented world. However, the education process should not be too narrow insofar as it detracts from the developments that science and engineering can provide for society. In the past, a high proportion of scientists and engineers with postgraduate degrees sought employment within academia. Furthermore, with the modern influx of postgraduate scientists and engineers working in nonacademic environments, the teaching process (including postgraduate education) must impart a broader range of skills as well as the willingness to participate in teamwork.

The result of these changes would be to emphasize technical proficiency as well as adaptability and versatility. To do this, both graduate and undergraduate programs must provide a broader exposure to experiences desired not only by academic employers but also by nonuniversity employers. Universities should offer more meaningful career information and guidance to students so that they can make well-informed decisions in planning their academic preferences and ensuing professional careers. Graduate education should prepare students for an increasingly interdisciplinary, collaborative, and competitive employment market and should not be viewed only as a byproduct of immersion in an academic research experience.

The science and engineering graduate-education enterprise should ensure a supply of preuniversity and university teachers, of university faculty, and of researchers in academic, government, and industrial laboratories. It should meet the expanding need for advanced scientists and engineers in careers outside of academic research and it should offer a diverse vision of education and employment that prepares future generations of students to strive for careers in practical (hands-on) science and technology.

This book focuses on the structure of the current education systems and follows the path of the primary objective of any educational institution, which should be the education of science and engineering students for careers in the academic and nonacademic worlds.

**James G. Speight**
*Laramie, Wyoming*

# Author

 **Dr. James G. Speight**, who has doctorate degrees in chemistry, geological sciences, and petroleum engineering, is the author of more than 60 books in petroleum science, petroleum engineering, and environmental sciences. He has served as adjunct professor in the Department of Chemical and Fuels Engineering at the University of Utah and in the Departments of Chemistry and Chemical and Petroleum Engineering at the University of Wyoming. In addition he has been a visiting professor in chemical engineering at the following universities: the University of Missouri-Columbia, the Technical University of Denmark, and the University of Trinidad and Tobago.

As a result of his work, Dr. Speight has been honored as the recipient of the following awards:

- Diploma of Honor, United States National Petroleum Engineering Society. For outstanding contributions to the petroleum industry. 1995
- Gold Medal of the Russian Academy of Sciences. For outstanding work in the area of petroleum science. 1996
- Einstein Medal of the Russian Academy of Sciences. In recognition of outstanding contributions and service in the field of geologic sciences. 2001
- Gold Medal—Scientists without Frontiers, Russian Academy of Sciences. In recognition of his continuous encouragement of scientists to work together across international borders. 2005
- Methanex Distinguished Professor, University of Trinidad and Tobago. In recognition of excellence in research. 2006
- Gold Medal—Giants of Science and Engineering, Russian Academy of Sciences. In recognition of continued excellence in science and engineering. 2006

# 1 Scientists and Engineers

## 1.1 INTRODUCTION

Science and engineering are major educational disciplines that, when put into practice, have resulted in advancements of the quality of life. Furthermore, the challenges of science and engineering have driven exploration and discovery for millennia, at least since 4000 BC; this was the time of the early Sumerians. They were the first recorded humans to use petroleum in a natural or modified form, and that use will continue into the foreseeable future (Speight, 2011, 2014). However, the future does raise the specter of challenges such as energy generation and protection of the environment—to mention only two such challenges—which also demand solutions that require the knowledge derived from science and engineering. Facing these challenges will be no small task and the education of young people through schools, preuniversity institutions, and universities will have to adapt to accommodate the needs of the future. Following such a path of adaptation will require a constructive and firm examination of the qualities of teaching, from the teacher in a school to the professor in a university.

Briefly and by way of definition, and to avoid any confusion, the use of the word *university* in this book includes all institutes of higher education beyond the high-school level. In conjunction with this definition, the use of the word *teacher* in this book refers to a person teaching at the preliminary-school and high-school levels of education, while the word *professor* includes all persons teaching at the post-high-school level of education—even though in the usual sense the title professor is given, in particular, to those persons who teach in a university. On the other hand, the term *teaching assistant* (frequently referred to as *TA*) as commonly used in a university is excluded from the definition of *professor* and, because it deserves extra discussion, is dealt with in more detail in a later chapter (Chapter 2).

Continuing with the historical aspects of teaching science and engineering, the products of science and technology have become more central and essential to society since the start of the Renaissance in the fifteenth century. A background in science and engineering has become essential to students moving into technical careers, as evidenced by the start of the Industrial Revolution in the late seventeenth century, whichresulted in the varied scientific and engineering inventions of the Victorian age. Particularly relevant in this respect are the inventions of Sir Humphry Davy (Fullmer, 1969, 2000) and Sir Michael Faraday. Faraday, although without a formal education, was educated by working for a book printer—as the printer's assistant, he printed and bound the books during the day and read them at night—and then he worked for Davy, who often offered challenges to Faraday through laboratory experimentation (Thomas, 1991; Russell, 2000; Hamilton, 2004). There are not many persons (if any) in the modern world who could repeat Faraday's educational achievements and, as a result, a degree in science and/or engineering is more fundamental to modern life than many previously sought-after arts degrees. The contributions of scientists and engineers have been allowed (in fact, it was necessary and, fortunately, unstoppable) to extend beyond laboratory research and development and reach into the realms of teaching, business, industry, and government.

Following on from the degree-bearing graduates and nondegree entrepreneurs of the nineteenth century (men such as—alphabetically and without favoritism— Isambard K. Brunel, Andrew Carnegie, Thomas A. Edison, J. Pierpont Morgan, John D. Rockefeller, and George Stephenson), a new breed of entrepreneurs has emerged—holders of various degrees from baccalaureate to doctoral degrees in science or engineering—and formed research companies, manufacturing companies, and consulting companies. They have also managed businesses, practiced law, formulated policy, and even run successfully for political office. In fact, within recent and unforgettable memory, Margaret Thatcher—holder of a baccalaureate in chemistry—became prime minister of Great Britain (from her election in 1979 to her retirement in 1990).

However, it is essential to remember that science-oriented and engineering-oriented students are not all alike any more than all artists or all politicians are alike, although the likelihood of there being another Margaret Thatcher or even a Steven Jobs should not be discounted. Success achieved by going where particular interests lead—performing a task or service that a person enjoys and getting paid for it—is half of the battle of life. Being exhilarated by the challenge of a new problem or by the complexity of bringing beneficial changes to the natural world is the *raison d'être* for any would-be or well-intentioned scientist or engineer. To the true *aficionado*, science and engineering provide the concepts that are necessary (and the means by which) to achieve personal goals which often mature into meaningful innovations.

Students in high school as well as young men and women in universities seek a career in science or engineering for a wide variety of reasons. As their experience progresses, some students evolve toward a technical career in science or engineering because these students are curious about the workings of the world and they have specific goals in mind—such as (1) creating new products, (2) decreasing pollution, or (3) working in other areas such as medical research and biomedical engineering. Other students may motivated by the theoretical aspects of science and engineering and move to the formulation of theories that, once proven and accepted, will assist their more practical hands-on colleagues to develop new inventions. These students, whatever their initial preference and once they have graduated, teach in various institutions from elementary school up to the university level, or they choose to provide counsel (hopefully wise counsel) or shape public policies on issues of relevance to science or engineering. Still other students use their education to work in developing countries where they may be engaged in educating indigenous students as well as helping to create and contribute to building a modern infrastructure.

Each of these motivations is legitimate as well as being valuable and commendable, and each follows naturally from an education in science and engineering. Scientists and engineers have specialized skills vital to national and international well-being that are manifested in teaching, basic research, applied research, and innovation

(Chapter 2). However, the need for teachers (in high schools) and professors (in universities) to focus on *teaching* is real and should not be taken lightly. It is the personal goals of the student along with capable mentoring by teachers and professors that guide the students through (what might seem to the student to be) the academic morass (Chapter 2) (NAS, 1997). The student needs to be guided along a path to help him/her to decide on the academic degree that is most appropriate and the type of teaching that must be applied to help the student reach his/her goals.

The first-step baccalaureate degree in science or engineering may provide entry into a satisfying career in a variety of positions. On the other hand, some students may find that the first-step advanced degree (a master's degree) may render them admirably equipped for a professional career. For other students, who are more inclined toward conducting research and/or teaching at the university level, a doctorate in a chosen discipline or subdiscipline is the more obvious or relevant choice (Chapter 2). However, it must be recognized that spending the required amount of time at a university may lead to an advanced (postbaccalaureate) degree but no degree in science or engineering—whatever the level—can guarantee lifetime employment in the highly competitive and ever-changing world. Like professionals in many other fields, the graduate might still have to change jobs and even career paths during professional life, perhaps more than once, with further education also being a necessary option.

Unfortunately, many students enter the science and engineering fields of study in a university without being aware of the trials and tribulations of such a life. It is often partway down the university track that the student discovers the stringent requirements and discipline that are involved in becoming a scientist or engineer. For some students, this may be an unmitigated career-path disaster but, assuming that the student recovers from this initial shock, s/he will probably want to understand, use, and explore science or engineering on a deeper level. That is the point at which the student, through capable and honest mentoring, turns to registration for an advanced degree (Chapters 2 and 6). At this time, the career course of the student may be set—if not set in concrete, at least written in soft concrete followed by concrete hardening and curing, by which time it might be too late to change paths. If the student has been mentored by being offered honest and unbiased guidance during the baccalaureate years (Chapters 2, 3 and 6) there would have been no surprises and the student may have chosen the right path. Moreover, graduate study for a master's degree or doctorate is mentally, physically, and emotionally demanding—anyone who thinks otherwise should make an immediate appointment with his/her psychiatrist! But not everyone has the perseverance to complete several years of concentrated study—and this is no shame—but the experience of being engaged in scientific or engineering work can be extremely exhilarating for those students with sufficient interest and determination (*true grit!*).

Thus, it is the responsibility of the education system to prepare the student(s) for the pitfalls that occur, to resist these pitfalls, and to continue on the straight and narrow path to scientific and engineering professionalism. After all, as the graduates progress through professional life, many will mirror the behavior of their respective mentors, be it ethical behavior or (unfortunately in some instances) unethical behavior.

Having now set the stage for the educational process and the issues that are likely to arise (or certainly that need serious consideration) as young people engage in these processes, it is the purpose of this book to help the reader understand that the educational path to a degree in science or engineering that can lead to a profession in those disciplines,, no matter what the level, can be a rigorous but enjoyable and fulfilling experience.

## 1.2  HISTORICAL ASPECTS OF SCIENCE AND ENGINEERING

The history of science and engineering is marked by a series of innovations in technology and applications of these innovations that have, fortunately for the most part, tended to be complementary (Singer et al., 1954; Landels, 1978; Hedrick, 2009). Innovations in technology lead to discoveries that are, in turn, advanced by other discoveries and inspire new possibilities and approaches to long-standing science and engineering issues. Therefore, investing in the education of would-be scientists and engineers is critical to ensuring technological advancement, but it is the quality of the education in science and engineering that is most important. In fact, the primary objectives of these scientific and engineering professionals are to create and develop innovative research concepts that (once the concept has moved to practical application) can be used to solve problems for organizations that will ultimately benefit the population at large. But again (and most of all), the high quality of education is the key component of technological advancement (Stensaker, 2013).

### 1.2.1  SCIENCE

The early beginnings of science are lost in the mists of time but (moving ahead to recorded history) in the later medieval period, as science in Byzantium and the Islamic world waned, Western Europeans began collecting ancient texts from the Mediterranean, not only in Latin, but also in Greek, Arabic, Hebrew, and some in Aramaic. Knowledge of ancient researchers such as Aristotle, Ptolemy, and Euclid were read with renewed interest in diverse aspects of natural phenomena while men such as Francis Bacon in England argued for more experimental science (Robinson, 2012). His attitude (and this might be borne in mind by readers) and the attitude of men like him was simple and to the point: "Do not spend time discussing and arguing about the number of teeth a horse might have; go, look, and count"—these are not the precise words of Francis Bacon but his thoughts put into words by the author.

By the late Middle Ages, especially in Italy, there was an influx of Greek texts and scholars from the collapsing Byzantine Empire. At this time, Copernicus formulated a heliocentric model of the solar system unlike the geocentric model of the Almagest—a commendable worth-reading astronomical manual written by Ptolemy (Claudius Ptolemaeus of Alexandria) in approximately AD 150. This work served as the basic guide for Islamic and European astronomers until the beginning of the seventeenth century. Subsequently, the Greek version of the Almagest was circulated widely in Europe, although the Latin translations from Arabic continued to be more popular and plentiful. However, all aspects of scholasticism were criticized severely (most were deemed heretical by the Pope and the Church) in the fifteenth, sixteenth, and seventeenth centuries—one author who was notoriously persecuted was Galileo Galilei, who made innovative use of experiment and mathematics (Pasachoff and Pasachoff, 2012). He was commissioned by Pope Urban VIII to write about the Copernican system, but after the work (*Dialogue Concerning the Two Chief World Systems*, also called *The Dialogs*) had been completed, the pope took offense and the persecution of Galileo began. Under threat (also under possible of application of torture) Galileo was forced to recant and the work was burned—although some of the original books have survived and the whereabouts of one copy of *The Dialogs* is known to the author.

Meanwhile, in Northern Europe, the new technology of the printing press was moving ahead (thanks to Johannes Gutenberg who introduced then applied the Chinese principles of movable type—along with other goods and ideas via the traffic on the Silk Road) at a rapid pace, and it was used to publish many books previously kept under lock and key (to which the Church had given the title "heretical texts" because they disagreed with church dogma). These books were given a wide distribution—this was a time when priests no longer had the sole privilege of being able to read. Many texts promoted the idea that science and engineering should study the laws of nature and this new modern science and engineering began to describe and formulate the physical laws of nature (the nontheological laws of nature). It was also argued by some authors (Roger Bacon being prominent among them) that science and engineering should aim for the first time at practical inventions for the improvement of mankind.

In the later seventeenth century and in the eighteenth century, new knowledge led to rapid scientific advance and the successful development of a new type of natural science, mathematical science, methodically experimental, and the need to be deliberately innovative. It is during this period that the word *science* gradually became more commonly used to refer to a *type of pursuit* of a type of knowledge, especially knowledge of nature—coming close in meaning to the term "natural philosophy." This led to the classification of the fields of science along two major paths: (1) natural sciences, which study natural phenomena, including the physical and biological life; and (2) social sciences, which study human behavior and societies. There are, presumably, purists who might otherwise wish to redefine the two categories of science. Be that as it may, in the sciences, whatever the subdiscipline, knowledge must be based on observable (and reproducible) phenomena as well as being experimentally verifiable (as a test of validity) by other research personnel working under exactly the same (laboratory or field) conditions.

Historically, the universities provided educational opportunities for young people but, sadly, only for the children of the wealthy. Furthermore, in medieval Europe, by this time there were only four faculties in a university: (1) theology, (2) medicine, (3) jurisprudence, and (4) arts, with the arts faculty having a somewhat lower status than the first three. Fortunately, modern universities have evolved from the disciplines of the mid-to-late nineteenth century as the traditional medieval curriculum changed and was supplemented by nonclassical languages and literature—Latin was no longer the language of learning—as well as by science and engineering disciplines. At the same time, the various subdisciplines of science and engineering had already been initiated and were evolving as separate subject areas.

In the early nineteenth century, science and philosophy (also called *natural philosophy*, referring to the study of nature and natural phenomena, whereas English speakers most typically referred to the study of the human mind as *moral philosophy*)—were considered to be synonymous. On the other hand, engineering was the type of engineering (i.e., civil engineering and military engineering) that had existed since (even before) the time of the Romans who seemed (fortunately) to have an ingrained DNA-based instinct to build, among other things, roads and aqueducts. As examples of ancient engineering projects, the ziggurat of Ur and the Egyptian pyramids spring to mind with the added involvement of an early form of chemistry using bitumen (derived from petroleum) as mastic when needed (Speight, 2014; J. G. Speight, Personal observations made at the remains of the cities of Nineveh, Calah, Dur Sharrukin, Babylon, and Ur, 1978). As the nineteenth century progressed, science and engineering continued to evolve into different disciplines and then into their current subdisciplines, which are now known collectively as *technology*. Furthermore, during the nineteenth century, the influence of and interest in science was increased in the various universities to (supposedly) match the influence and interest of engineering, which was enjoying a high degree of social status due to the successes achieved during the formidable Industrial Revolution; these had focused on issues related to the development of various projects such as the development of railways and other engineering projects such as bridge building.

As a result of the increased influence of science in the universities, European (specifically German) academics categorized the study of nature into four subcategories: (1) physics, (2) chemistry, (3) biology, and (4) geology (Grau, 1988). To complement the movement of science into academia, the academicians also organized the university science administrative units in the same way. With the passage of time, other universities throughout the world eventually followed the lead of the European universities and used the same categories to establish science departments that were also focused on physics, chemistry, biology, and geology.

As the nineteenth century progressed, science and engineering (which had commenced to evolve as two separate umbrella catch-all disciplines) began to separate from philosophy, though science (with engineering) often retained a very broad meaning and philosophy still found a way into both disciplines. In many cases, "philosophy" stood for reliable knowledge about any topic and was generally linked to a set of well-defined laws—not just of the laws of nature but laws that could be applied to any phenomenon that could be described by chemical or physical laws. In addition, over the course of the nineteenth century there was an increased tendency to associate science with the natural world (i.e., the nonhuman world) and engineering with Roman-type construction projects. At the end of the nineteenth century and into the beginning of the twentieth century, the evolution of science and engineering as separate disciplines tended to eject philosophy from the science and engineering umbrella, which left the study of human thought and human society (what would come to be called *social science*) in scientific no-man's land or certainly in scientific limbo.

Furthermore, throughout the nineteenth century (predominantly the Victorian Age—1837–1901), many English speakers were increasingly differentiating science from engineering and both disciplines from all other forms of knowledge. For instance, the now-familiar expression *scientific method* (i.e., the experimental procedures used to explain the events of nature in a reproducible way and to use the experimental data to make meaningful predictions) was almost unused during the early part of the nineteenth century and only came into common use after the 1870s, though there was not always total agreement about the precise definition of the scientific method and what it involved. Similarly, separation of scientists and engineers as a special group of technology-oriented people who were involved in scientific and engineering projects did not always emphasize the attributes of their respective education systems. And then there was the language of communication—modern listeners to scientists and engineers may often wonder what language is used to transmit thoughts and ideas to each other and between the members of each discipline. In fact, the language used to communicate scientific and engineering thoughts, ideas, and concepts, as well as experimental data, tended to depict science and engineering as disciplines that were separate and distinguishable and caused a dilemma of communication within the education systems of the technologically oriented university departments in many countries.

At the same time, as the nineteenth century drew to a close, the school science and engineering curricula were further defined and developed in Europe and North America. On the one hand, the science curricula were being reorganized to follow the administrative units of physics, chemistry, biology, and geology as taught in most universities. On the other hand, engineering—as it often still is, much to the detriment of the discipline and the student(s)— was lodged (even hidden) within the mathematics curriculum with little (or no) attention paid to engineering as a formal discipline of learning. In fact, the typical high-school engineering curriculum (if there is such a curriculum) continued as an offshoot of the mathematics curriculum, which made it impossible to deal with or even to acknowledge the various engineering subdisciplines. In fact, some universities found it necessary to offer preengineering courses to acclimatize the students to the realities of engineering—a practice that, unfortunately, is still necessary in many universities of the twenty-first century. The attitude that "this is how we have always done it and we will not change our system" pervades many schools and universities, very much to the detriment of first-year (engineering) students who wish to enroll in engineering departments as engineering students and not as mathematical students.

### 1.2.2 ENGINEERING

The history of the concept of engineering stems from the earliest times when the use of the pulley, the lever, or the wheel to build houses, roads, and bridges was common. The exact origin and historical development of a linguistic form of the word *engineer* is lost in the mists of time, but as the word is currently used, it indicates a person occupationally connected with the study, design, and implementation of engines. The word engine derives from the Latin *ingenium* meaning *innate quality*, especially involving applying mental power to the task or purpose of an innovation. Hence, an engineer is, essentially, a person who has the ability to develop useful or practical inventions from laboratory concepts. The current definition would also include the ability of the engineer to initiate innovative concepts—that is, the engineer is also a potential inventor.

Works spread throughout the world such as (not listed here in any preferential or chronological order) the Acropolis and the Parthenon in Greece; the Roman aqueducts; Via Appia (one of the main highways in the Roman Empire) and the Coliseum; the Hanging Gardens of Babylon; the Pharos of Alexandria; the pyramids in Egypt; the spiritual city of Teotihuacán and the cities and pyramids of the Mayan, Inca, and Aztec Empires; and the Great Wall of China, among many other works, represent the ingenuity and skill of the ancient civil and military engineers. In fact, through time, the Silk Road that started its existence as a caravan track became a more formal roadway that was maintained by the various countries through which it traversed.

On a human note, the earliest civil engineer known by name is Imhotep—an ancient Egyptian medical doctor, architect, high priest, scribe, and vizier (chief advisor) to King Djoser (Netjenkhet Djoser, the 2nd King of Egypt's 3rd Dynasty). As an official of the Pharaoh Djoser, it is very likely that Imhotep designed and supervised the construction of the Step Pyramid (at Saqqara in Egypt) during the period 2630–2611 BC. He may also have been responsible for the first known use of columns in the architecture of palaces and temples.

Engineering in Roman times was originally divided into (1) military engineering, which included construction of fortifications as well as military war machines, and roads to move armies; and (2) civil engineering, involved in nonmilitary projects such as aqueducts, bridge construction, and building construction—such as the Pantheonand the Baths of Caracalla, both in Rome. From another perspective, another meaning of engineer, dating from 1325 and considered to be obsolete by some linguists or verbal purists, is a constructor of military engines, such as Leonardo da Vinci (April 15, 1452–May 2, 1519).

The first electrical engineer (who is noted as such) is considered to be William Gilbert, with his 1600 publication of *De Magnete* and who was the originator of the term "electricity." The first steam engine was built in 1698 by mechanical engineer Thomas Savery—the development of the steam engine gave rise to the Industrial Revolution in the coming decades, allowing for the beginnings of mass production of materials and equipment. Other electrical engineers worthy of note (and recognizable names) include: Alessandro Volta, Michael Faraday, and Georg Ohm. The work of James Maxwell and Heinrich Hertz in the late nineteenth century eventually gave rise to the field of electronics—the later inventions of the vacuum tube and the transistor further accelerated the development of electronics to such an extent that electrical and electronics engineers currently outnumber their university colleagues in many other engineering disciplines and subdisciplines.

The inventions of the English inventor Thomas Savery (1650–1715) and the Scottish engineer James Watt (1736–1819) gave rise to a version of engineering that eventually came to be known as mechanical engineering. The development of specialized machines and their maintenance tools during the Industrial Revolution led to the rapid growth and development of mechanical engineering both in its birthplace (Britain) as well as in many other countries, most of which were under British influence, with the exception of the United States. However, even though in its modern form mechanical engineering originated in Britain, the origins of mechanical engineering can be traced to early antiquity where ingenuous machines were developed both for military and civilian purposes. For example, the Antikythera mechanism, the earliest known model of a mechanical computer in history, and the mechanical inventions of Archimedes are examples of early mechanical engineering. Some of the inventions of Archimedes, as well as the Antikythera mechanism, required a knowledge of the intricacies of differential gearing or epicyclic gearing (two gears mounted so that the center of one gear revolves around the center of the other), two key principles in machine theory that helped design the gear trains of the Industrial Revolution and are still widely used in diverse fields such as robotics and automotive engineering (Wright, 2005).

## 1.3   DEFINITIONS

Science and engineering are the disciplines of acquiring and applying scientific and engineering knowledge. On the one hand, science is in fact a branch of knowledge or study dealing with facts learned through experiments and observation and systematically arranged in the form of general laws of the natural world. Science is further subdivided into particular areas of scientific study such as biology, chemistry, and physics, which are further subdivided into various subdisciplines or subcategories. On the other hand, engineering is the art or science of making a practical application of the knowledge of pure science (such as chemistry or physics) or that gained from laboratory-based engineering projects, as in the construction of chemical plants, mines, bridges, buildings, mines, ships, and the like.

In addition, the word "science" is inclusive of the life sciences, physical sciences, mathematics, and social science (the study of human society and social relationships) as well as of political science (also a study of human society and social relationships but with a different aspect to social science, and which some observers consider to be not based on any form of logic). The word "engineering" refers to a field of endeavor that includes all subdisciplines such as civil engineering, mechanical engineering, electrical engineering, petroleum engineering, computer engineering, and environmental engineering. There may be other recently developed subdisciplines of science and engineering that are specific to a particular university or to a particular country and these are implied or understood though not directly expressed here, and are nevertheless included without the disrespect of not being mentioned.

By the onset of the twentieth century, the modern notions of science and engineering as special disciplines that offered studies of the technological aspects of the outside (nonacademic) world were essentially in place and were being practiced by a distinct group of professionals and pursued through knowledge and application of practical methods. Indeed, as the twentieth century progressed and reached maturity, the cooperation and links between science and engineering (much to the benefit of technology) grew stronger. In fact, it is arguable that by the end of the twentieth century and the dawn of the twenty-first century, the term "science and engineering" was being eclipsed by the new term *technology*, as a term of public attention and interest.

In fact, in the twenty-first century, there is an increasing tendency within the hallowed halls of academia to incorporate fields of study that are created by further subdivision of the traditional scientific and engineering disciplines, even to the stage where new disciplines (more specifically subdisciplines) have been created. Furthermore, the disciplines are further divided (computer engineering is one such example), defined, and then recognized by the learned journals in which scientific and engineering research is published as well as by the societies and academic departments or faculty organization to which the respective practitioners belong. However, caution is advised since the creation of further subdivisions (or even sub-subdivisions) of science and engineering may only serve to add confusion to the already confused and much crowded system of subdivisions.

Thus, a scientific or engineering discipline is a particular branch (identified by name) of science or engineering. There is, however, more to the definition than these words or simple meanings would indicate, but the definition will suffice for the purposes of this book. Moreover, throughout this book, the words "scientist" and "engineer" refer to persons who wish to pursue or those who have already attained at least a baccalaureate degree (or equivalent qualification) in science or engineering, respectively, as well as those persons who have also been awarded an advanced postbaccalaureate degree in science or engineering.

### 1.3.1 SCIENTIFIC DISCIPLINES

Science (Latin: *scientia*—knowledge), in the broadest sense, refers to any systematic knowledge or practice and also refers to a system of acquiring knowledge based on the scientific method, as well as to the organized body of knowledge gained through scientific research. However, the broad discipline of science encompasses a range of specialized subdisciplines that focus on the issues associated with using a specific type of science or following a specific line of research as well as developing a specific kind of product. In addition, mathematics, which is sometimes classified within a third group of scientific disciplines called *formal science*, has both similarities and differences with the subcategories of science and engineering. It is similar to empirical sciences in that it involves an objective, careful, and systematic study of an area of knowledge but it does differ because the mathematical method of verifying knowledge uses *a priori* methods rather than empirical methods.

A person who practices science is, therefore, a scientist, and those licensed to practice science by a society or licensing organization (Chapter 3) have more formal additional designations such as Chartered Chemist (Royal Society of Chemistry,* London, United Kingdom). In some societies, such designations may be equivalent to a baccalaureate degree and give the holder the right to practice his/her profession with a high level of knowledge as well as acknowledged hands-on capability.

### 1.3.2 ENGINEERING DISCIPLINES

In all senses of the word, "engineering" refers to the discipline of acquiring and applying technical knowledge to the design, analysis, and/or construction of equipment or works for practical purposes. Alternatively, engineering is also the creative application of scientific principles (such as in the subdiscipline chemical engineering) (1) to design or develop structures, machines, apparatus, or manufacturing processes, or works utilizing them singly or in combination; or (2) to construct or operate the same with full cognizance of their design; or (3) to forecast the behavior of constructed works under specific

---

* http://www.rsc.org/Membership/Qualifications/CharteredStatus/cchem.asp (accessed August 5, 2013).

operating conditions with respect to the intended function, economics of operation, and safety to life and property (ECPD, 1941). In fact, one branch of engineering—civil engineering—is a field that deals with buildings, bridges, dams, roads, and other structures. Civil engineers (who evolved from the ancient military engineers) plan, design, and supervise the construction of facilities such as high-rise buildings, airports, water-treatment centers, and sanitation plants. Civil engineers will be needed to assist in the design of the special rail beds for the magnetic levitation trains of the future.

With the rise of engineering as a profession during the nineteenth century the term engineering became more generally applied to fields in which mathematics and science were applied to industrial uses. Similarly, in addition to military and civil engineering the fields then known as the mechanic arts became incorporated into engineering curricula.

Chemical engineering is that branch of engineering that processes raw material by chemical, physical, or biological means into different products (Shallcross, 2005). Modern chemical engineers work in a range of industries, designing, building, and operating processes that transform crude oil into gasoline and plastics, produce a range of specialty products from raw milk, and capture carbon from the smoke stacks of coal-fired power stations. Chemical engineers need to understand the principles of a range of topics, including biological processes, control of processes, fluid flow, heat transfer, mass transfer, material balances, momentum transport, process dynamics, process equipment design, reaction processes, safety, separation processes, solids handling, sustainability, and thermodynamics. As well as possessing technical knowledge and skills, engineers are problem solvers able to break complex problems into more manageable tasks. As with many other disciplines, chemical engineers must be effective communicators, be able to work in teams, and have a sound understanding of management practices and process economics.

Chemical engineering, like its counterpart mechanical engineering, developed in the nineteenth century during the Industrial Revolution. Industrial-scale manufacturing demanded new materials and new processes. By 1880 the need for large-scale production of chemicals was such that a new industry was created—the petrochemical industry—that was dedicated to the development and large-scale manufacturing of chemicals. The role of the chemical engineer was the design of these chemical plants and processes such as processing and treating liquids and gases. Many chemical engineers work with petroleum and plastics, although both of these are the subject of independent disciplines. The subdiscipline environmental engineering also applies to certain areas of chemical engineering, such as pollution control.

Aeronautical engineering deals with aircraft design, while aerospace engineering is a more modern term that expands the reach envelope of the discipline by including the design of space craft. The origins of aeronautical engineering can be traced back to the aviation pioneers such as the Wright brothers in the late nineteenth century and early twentieth century, although the work of Sir George Cayley has recently been dated as being from the last decade of the eighteenth century.

The Wright brothers, Orville (August 19, 1871–January 30, 1948) and Wilbur (April 16, 1867–May 30, 1912), were inventors and aviation pioneers who are credited with inventing and building the world's first successful airplane and making the first controlled, powered, and sustained heavier-than-air human flight on December 17, 1903. From 1905 to 1907, the brothers developed the first practical fixed-wing aircraft.

Although not the first to build and fly experimental aircraft, the Wright brothers were the first to invent aircraft controls that made fixed-wing powered flight possible. From these early pioneering efforts, knowledge of aeronautical engineering was developed with concepts imported from other branches of engineering. In fact, little more than a decade after the success of the Wright brothers, the 1920s saw the development of aeronautical engineering through the need for military aircraft in World War I.

Just as scientific disciplines can be said to span the alphabet, engineering disciplines also are manyfold and offer a wide choice of options. New subdisciplines such as bioengineering, which combines biology and engineering, are thriving. Bioengineers work closely with biologists and medical doctors to develop medical instruments, artificial organs, and prosthetic devices. In addition, the relatively new subdiscipline of computer engineering (developed over the past 4–5 decades) deals with all aspects of computer systems including design, construction, and operation. Some computer engineers specialize in areas like digital systems, operating systems, computer networks, and software. For example, computer engineering (or computer systems engineering) encompasses broad areas of both electrical engineering and computer science. Computer engineers are engineers who have training in the areas of software design and hardware–software integration. In turn, they focus less on power electronics and physics than do electrical engineers. Some areas in which computer engineers are involved are software development, hardware (firmware/software) integration, circuit design, and system-level design and integration.

Thus, in summary, a person who practices engineering is an engineer, and those licensed to do so have formal designations such as Professional Engineer (National Society of Professional Engineers,* Alexandria, Virginia), Chartered Engineer (Engineering Council, London, United Kingdom), or Incorporated Engineer (Institution of Engineering and Technology, Stevenage, United Kingdom). The broad discipline of engineering encompasses a wide range of specialized subdisciplines that focus on the issues associated with developing specific kinds of products or use specific types of engineering knowledge and practice.

## 1.4   THE CONCEPT OF A PROFESSION

The current definition for the scientific and engineering subdisciplines indicates that these areas of study are disciplines and branches of knowledge that are taught at the university level, often with the associated and relevant postbaccalaureate research programs. Persons educated as scientists and engineers are trained to provide concepts and develop original ideas that are brought to fruition in teaching, industry, business, and government. Graduate students, if tutored correctly, can move well beyond the thinking of their respective professor and create new levels of scientific and engineering thought. The student learns from the professor, but the professor, if s/he will ever care (or dare) to admit it, also learns from the student.

In simple terms, a profession is any occupation that provides a means by which a person with a professional qualification can earn a living. In the sense intended here and without any disrespect to persons working in trades that are not included in the

---

* http://www.nspe.org/index.html (accessed August 5, 2013).

definition of profession, the scientific and engineering professions are those forms of work involving advanced expertise, self-regulation, and ethical behavior (Martin and Schinzinger, 2005). Furthermore, scientific and engineering professionals play a major role in setting standards for admission to the profession by (1) drafting codes of ethics, (2) enforcing standards of conduct, and (3) representing the profession to others. Professionals should maintain high ethical standards, and to do so brings with it the recognition traditionally associated with the word profession. Thus, in the present context, a profession is founded upon specialized educational training, the purpose of which is to provide objective service to others, for a direct and definite compensation, without expecting any other form of gain (Lammas and Garcia, 2009; Olgiati, 2010).

## 1.4.1 SELECTING A PROFESSION

In spite of the varied definition of the word, there is considerable agreement about defining the characteristic features of a profession. Typically a professional is a member of a professional association, (Chapter 3), has received institutionalized training (Chapter 2) and licensing (Chapter 3), and adheres to a code of ethics (Chapter 3). Members of a profession have also been defined as "workers whose qualities of detachment, autonomy, and group allegiance are more extensive than those found among other groups…their attributes include a high degree of systematic knowledge; strong community orientation and loyalty; self-regulation; and a system of rewards defined and administered by the community of workers" (Brown, 1992).

Originally, any regulation of a profession was self-regulation through a professional association. However, with the growing role of government, statutory bodies have increasingly taken on this role and the members are appointed either by the profession or (increasingly) by government. Proposals for the introduction or enhancement of statutory regulations may be welcomed by a profession as protecting clients and enhancing the quality of service, or the regulations may be resisted because they are seen to limit the freedom of the members to innovate or to practice in a manner which, in their professional or personal judgment, the members consider best. Besides regulating access to a profession, professional bodies may set examinations of competence and enforce adherence to an ethical code (Chapter 3).

Typically, individuals are required by law (sometimes by peer pressure) to be qualified by a local professional body before they are permitted to practice in that profession. However, in some countries, individuals may not be required by law (or peer pressure) to be qualified by a professional association in order to practice the profession. In such cases, qualification by the professional association is still considered to be a prerequisite to practice, as most employers and clients stipulate that the individual must hold such qualifications before s/he is hired. However, profession associations tend to be autonomous, which means they have a high degree of control of their own affairs, and members of the association usually have the right to exercise their individual professional judgment. In such cases, the scientist or engineer may find that by membership in a professional association, s/he has a measure of protection against liability for professional negligence, but there is no professional association that offers protection to the members who practice the profession in a negligent manner—that is, there is no form of protection for gross negligence.

Part of the continuing education of scientists and engineers as they move into the workforce is whether to be (1) a member of an academic department, (2) a government scientist or engineer, or (3) in the commercial world. Either appointment could require that the scientist or engineer work as part of a team (Chapter 6). A team is a collection of scientists and/or engineers with the same goal in mind—to complete the project successfully within the allotted time and within the allotted budget. Thus, being a member of a team requires alignment of thinking to focus on working in a cooperative (usually selfless) manner and toward a specific business purpose, and involves: (1) the sacrifice of personal goals, (2) the sharing of rewards, (3) the sharing of blame and punishments, (4) uniformity of thought, and (5) suppression of personal opinions, none of which are palatable to many scientists and engineers.

Businesses and other organizations often go to the effort of coordinating team-building events in an attempt to encourage scientists and/or engineers to work as a team rather than as individuals. Universities are less conscious of teamwork, where every professor is his/her own independent island (or monarchy) with much (usually *all*) of the authority but a distinct unwillingness to accept any responsibility for his/her actions. Furthermore, the problems that can arise from a single autonomous investigator working on a project can (but not always) be overcome by appointing a principal investigator to the project. This is typically the decision of the agency or corporation that sponsors and funds the research.

Once the graduate scientist or engineer has chosen his/her career path, an issue arises if the choice has been to go into (or to remain) in academia. The choice is related to the concept of *tenure*.

### 1.4.2 The Concept of Tenure

Any discussion related to the definition and discussion of the concept of a profession in science and engineering would be incomplete without acknowledgment of the concept of tenure. It is because of tenure that members of teaching faculty (in schools) and members of the academic faculty (in universities) receive frequent criticism from students and parents and come under equally frequent criticism from other sources, such as the news media. Tenure has typically been used (and defined) to provide school teachers and university professors with job protection (Smallwood, 2003; Batterbury, 2008).

It is believed that tenure originated in schools during the period of the late nineteenth century into the 1920s when teachers, especially female teachers (who dominated the teaching profession—unions for teachers had not been formed at that time), started fighting for their rights. Teachers came together to gain protections for such rights, which ensured they could not be fired without reason. Tenure in the universities may have raised its head shortly thereafter, or even before the awarding of tenure to teachers.

The concept of tenure was originally conceived to allow faculty members to continue teaching in the area of scholarship at the university that awarded the tenured position. In fact, tenure has become a feature of university employment as well as employment in a school. However, "tenure" has evolved over the decades to mean "job protection," and an appointment without tenure is not always a palatable option for many would-be teachers and would-be professors. Indeed, an appointment to a position where continued employment is dependent upon measured standards of performance is also not always palatable unless tenure goes with the appointment.

Many observers in the nonacademic world heartily (and loudly) disagree with the concept of tenure as they feel it tends to promote a high feeling of job security and removes the impetus to work conscientiously as well as the impetus to reduce any perceived workload. In other words, many observers feel that those with tenure have ultimate protection from accountability and the permanence of the appointment is unlikely to be affected (Sykes, 1988,; Glencorse, 2013). Tenure is a way of life for the scientist and engineer who joins a university faculty but not for the scientist or engineer in an industrial laboratory or in a government laboratory where promotion and job security are based on performance.

Many faculty members in schools and universities consider tenure to be an employment contract with no specific end date but there is no guarantee of tenure at the commencement of academic employment. Reviews for tenure consider many factors and there are many levels of review, and only the governing board (the board of trustees, the board of regents) supposedly can grant tenure (Holcomb et al., 1987). There is the misconception that once tenure is awarded, the appointment (from lower faculty status to upper faculty status) is a job for life, but tenure, in reality, means that the appointee of tenure is qualified to continue teaching at the university (McKenzie, 1996; Hill, 2014). Thus, for the wrong reasons, the common definition of a "tenured professor" is a faculty member who has been given a job for life at his/her university. Furthermore, it is also a general belief (actually it is a misbelief) that a tenured teacher or professor cannot (will not) be fired, except in extreme circumstances, such as committing a very serious crime, and even then, extenuating circumstances may arise from somewhere and for some (not always logical) reason. The holders of tenured positions can often (singly or collectively) rationalize a decision or action (however bad or wrong) seven ways from Sunday!

The actual activities of faculty members, how their activities are perceived, and how these activities are rewarded (salary increases are also issues in the academic world but are not the subject of this text) are varied. Nevertheless, all roads in academia lead (not to Rome but) to the concept of tenure. Thus, for those scientists and engineers who choose an academic career as their professional goal, the concept of tenure will arise or, perhaps, should have arisen before the decision to lean toward academia had been formulated. During the doctorate work, the student will have been exposed to the idea of tenured professors, as well as some who are not granted tenure. Therefore, this is an opportune place to insert come comments on the concept of tenure.

The definition of tenure generally focuses on academic freedom in the United States (Hofstadter and Metzger, 1952; Conrad, 1993; AAUP, 2006; Andreescu, 2009; Robinson, 2013). In the United States, academic freedom means that a university can determine—on academic grounds—the following: (1) who may teach, (2) what may be taught, (3) how it should be taught, and (4) who may be admitted to study. It has also been decided in court (Stronach v. Virginia State University, civil action 3:07-CV-646-HEH, E.D. Va. Jan. 15, 2008) that professors have no academic freedom and *all* academic freedom resides with the university or college.

The most common definition of tenure is the status of a teacher or professor holding his/her position on a permanent basis without the need for periodic evaluation and contract renewal. Although the constitutions of various universities and related other documents clearly define tenure, the definition may not always be enforced in the event of wrongdoing a holder of tenure. Thus, it is not surprising that considerable confusion remains among the general public (and many colleagues in science and engineering) as to the meaning of tenure and the evaluation of performance for its achievement. In theory as well as in practice, tenure is not (or should not) be a guarantee of lifetime employment but it is looked upon by many faculty as just that.

Under the tenure systems adopted as internal policy by many universities, colleges, and schools—especially in the United States and Canada—tenure is associated with more senior job titles such as *associate professor* and *professor*. An assistant professor (or a nontenured teacher) will not be promoted to a tenured position without meeting the goals of the institution, often (though not always including) demonstrating a strong record of teaching, published research, project funding, academic visibility, and administrative service, with emphases being different across institutions (though often focused on research in universities). Typical academic protocols allow a limited period to establish such a record and limit the number of years that any faculty member can hold a junior appointment title such as *lecturer* or *assistant professor*—in the United States and Canada, the lecturer position is often below that of assistant professor, whereas in the older European universities the title lecturer was equivalent to the U.S./Canada associate professor. An institution may also offer other academic titles that are not time-limited such as *adjunct professor*, *research professor*, or *visiting professor* but these positions do not carry the possibility of moving up the academic ladder to be awarded and occupy a tenured position and, therefore, are often described as nontenure track appointments.

The process of earning tenure typically lasts approximately 6 years, during which the candidate's teaching, research, and service are evaluated by departmental tenured faculty. Annual evaluations (Section 1.6), which should be taken very seriously, are required but are not always performed with any degree of regularity or professionalism. At any time during the 6-year probationary period, the individual can be notified, supposedly with specific advance notice and with cause, that her/his contract will not be renewed. Whether or not unbiased professionalism (by the departmental tenured faculty and/or by department head/chair) is practiced in such cases is not always obvious or guaranteed.

However, each university should formulate an *official* institutional definition of adequate cause, specifically misconduct or incompetence. If this is not done or enforced by the institution, the fault must be laid upon the shoulders of the faculty members, who have the responsibility to participate fully in setting the standards and expectations for themselves and each of their colleagues. Unfortunately, this is a responsibility that is put into practice infrequently—in many cases it is not put into practice at all. The failure of faculty to shoulder and to exercise this responsibility is the main source of much of the frustration with the behavior of the faculty members and, hence, with schools and universities.

Tenure at many universities depends solely on research publications and research grants, even though there are statements in the official policies of the university that tenure depends on research, teaching, and service to the university or to the community (Boyer, 1990). Even articles in refereed journals and project funding may not count (for reasons unknown) toward tenure within some university departments. But assuming that all is well, at many universities the department chairperson sends forward the departmental recommendation on tenure. This is not always final, as there have been instances where the faculty voted unanimously to tenure an individual but the chairperson or the dean of the college sent forward a recommendation not to grant tenure to an individual with an opinion (not always written, often verbal) that is diametrically opposed to the recommendations of the faculty members who sat and deliberated as members of the tenure committee.

On the other hand, if everyone of consequence is in agreement, once tenure is granted, the newly tenured members becomes part of an operative system (an "old boys club" or an "old girls club") that is often disrespectful to those faculty members who do not have tenure, and those faculty members who have achieved tenure can (and often do) become averse to any form of teaching and/or research activity. The tenured faculty can resist necessary reforms by administrators whom they generally outlast—many administrators are often employed on time-limited contracts to be renewed or not to be renewed. Typically, the tenured faculty members also control academic appointments and such control contributes to the practice of hiring more of the same kind of person, thus propagating the system for eternity (Riley, 2011).

In the past, there have been calls for changes to the tenure system (Searle, 1971; Sykes, 1988) and it has been suggested that capable professors be given tenure much sooner than the standard 4–6 years so as not to hamper their classroom teaching. However, there was the accompanying suggestion that tenured professors be reviewed on a designated schedule (say, every 7 years) to help eliminate incompetent teachers who otherwise would find refuge in the tenure system (Searle, 1971; Derrida, 1988). It has also been suggested that tenure may have the effect of diminishing academic freedom among those seeking it—that they must appear to conform to the political or academic views of the field or the institution where they seek tenure, and thus tenure may have the opposite effect to its stated goal of diversifying free expression (Michaels, 2004; Robinson, 2013). Other criticisms include the *publish or perish* pressures in which faculty members publish papers that have very little to offer to the scholarly literature but which add up, one paper by one paper, to a creditable number in a specific time period.

In what should be normal (some would use the word *typical*, being unable to define normal in academia) circumstances, the final tenure evaluation is conducted by scholars from across the university community and includes assessment by individuals from other universities. The decision on tenure means that a collection or committee of scholars has judged the candidate to have excelled (or not excelled) in teaching, research, and service; and that it has (or has no) confidence in the individual's judgment, after which the candidate is welcomed into (or rejected from) membership of the academic community. On a practical level, the major difference between the untenured individual and the tenured faculty member is that the untenured individual can be removed from academia fairly easily while the tenured individual can only be removed (with some difficulty) from academia for adequate cause. Thus removal of either nontenured or tenured faculty typically involves building a case (personnel) file of wrongdoings or "no-doings" (failure to attend departmental meetings and refusal to teach courses). At the same time, the tenured miscreant may have been awarded annual/regular salary increases to keep him/her quiet. So the university may have difficulty in building a reliable and prosecutable case against the miscreant!

As a result, it is not surprising that critics have observed (and stated) that tenure makes it virtually impossible to fire bad/incompetent teachers and professors—getting rid of teachers or professors with tenure can involve years of review and bureaucratic hurdles, and can cost tens of thousands of dollars per teacher/professor (Garrett, 2013). Another issue is that teachers and professors are not properly (objectively) evaluated before being granted tenure. Many schools and universities have set the bar far too low for the achievement of tenure and it is no longer based on teacher or professor effectiveness, but instead on a certain length of service by the teacher/professor. With tenure being granted after only a few years, some teachers or professors may not have the chance to show their worth, or their ineptitude. That is a legitimate concern and a school principal or school board (university dean or president and the board of trustees) should not be afraid of advising a teacher (or professor, if a similar situation exists in a university) that teaching is not his/her profession (Garrett, 2013). In fact, it would be good governance to take such action (Salmi and Helms, 2013).

## 1.5  EVALUATION

Unfortunately, the processes by which scientists and engineers are evaluated in academia and in industry may be the most detrimental effects that can decrease the will of the scientist or engineer to perform at an adequate level. Many evaluators focus on the faults and errors made by the interviewee—it is easier to find fault than it is to offer praise and to see the benefits of someone's work. Most of all, the evaluation process should involve knowledge of the education of scientist and engineers as well as being able to speak their language—being evaluated by one or more members of the nontechnical personnel administration is not the way to proceed! However, there are other aspects to getting the best out of scientists and engineers, and these relate to the evaluation process.

In addition to the scientist or engineer who may not appear to fit the academic or company mold, the evaluation process may seem to focus on the "do-as-I-say" dictum of the immediate supervisor, department head, or academic senior colleagues. In academia, the additional dictum of *publish or perish* is also operative, insofar as "publish" has the standard academic meaning of "publish in recognized journals that we, the evaluation committee, say are recognized journals," irrespective of the knowledge of the members of the evaluation committee to the reputation of the journal—and we should not forget, the number of papers published in the time period from the last evaluation is an easy mathematical exercise for any supervisor. Relying on journal publications as the sole demonstration of scientific and/or engineering achievement is a sad state of affairs and needs to be a thoroughly reevaluated with the realization that *pro forma* evaluations should also be discontinued.

In the first instance (i.e., the "do-as-I-say" syndrome), the young professional may find that s/he is up against the proverbial brick wall. The supervisor/department head is all powerful and the scientist or engineer has little or no recourse for appeal. Equally, the do-as-I-say syndrome is also fraught with pot holes for the young scientist and engineer. Production of patentable work may also require acknowledgment of the supervisor and any other designated person as coauthors, which is also ruinous to the young scientist and engineer who had the original idea and brought it to experimental proof. Providing money and facilities for the project is not a criterion for coauthorship—some managers of government contracts may disagree with this thought as may some industry-based project managers. But these worthies are often absent from the scene when the work does not product patentable ideas and the project is terminated. Furthermore, blame is assigned to the young scientist and engineer while the supervisor who originally (perhaps half-heartedly) supported the concept and any other potential hangers-on seeking coauthorship of a successful idea have backed away and are not evident by any form of presence or support.

Furthermore, counting the number of publications in recognized journals (recognized by whom?) ignores (1) the nature of the journal, (2) the quality of a publication, as well as (3) the potential for benefit through ownership (by the university) of intellectual property. For example, publication of work in a patent followed by publication of the work in a conference proceedings are tangible means of conveying ideas and insight that relate to intellectual property. Obligating scientists and engineers to be evaluated without giving true credence to intellectual property is a handicap and is ruinous of a true method of evaluation.

It might be said that the real education of scientists and engineers involves evaluation of the work by the individual as well as their performance either through teaching activities in the classroom and/or through actions in the laboratory as part of a research team. In either case, the evaluation process must be objective and honest, with no bias toward the scientist or engineer or toward the outcome of the work. Furthermore, if the evaluation process involves an annual meeting between scientist/engineer and supervisor, the process should not involve a surprise at the end of the year—the annual performance evaluation should be a reaffirmation of the comments made during constant communication throughout the year.

Indeed, there are several scenarios that do not encourage either academic achievement or industrial achievement in science and engineering. Dealing with such issues is part of the education that all scientists and engineers should face in any year of their respective careers.

## 1.5.1 THE PROCESS

The scientific and engineering disciplines and subdisciplines are composed of educated and relatively young professionals who have the ability to apply themselves to the problems at hand, whether theoretical studies or experimental (laboratory) work. To the scientist and engineer, the outcome of this work that offers some form of gratification is (1) completion of a project and (2) publication of the data in a journal or similar medium for distribution to one's peers. The latter gives the scientist and engineer recognition for their work (Shaw and Despota, 2013).

However, the scientist and/or engineer may be required to establish that s/he has performed better insofar as the work can be established as better than the work of any predecessor and, if possible, offers an advance to the work of predecessors. Then the definition of *better* becomes an issue that can detract from the real issue—getting the best out of scientists and engineers. Put simply, "better" can mean (1) a process for producing an improved product, (2) completion of a project on time and under budget, or (3) anything in between these two limits. If these criteria are not applied objectively, the fundamental basis for deciding on how to reward achievement may be lost because of a complete misunderstanding of the nature of the work and its importance to the university or to the company. The final effect is that the contribution is lost in the midst of an argument focused on word definitions.

For the purposes of evaluating a scientist or engineer, there should be two critical objectives of an evaluation and these are (1) recognition of the significance of the contribution and any ensuing, and (2) the magnitude and significance of the impact of the contribution. During this process, if the young scientist or engineer perceives that his/her work is being recognized as meaningful, increased morale develops and the young professional is satisfied, for the moment! But the means by which the impact of the contribution can be assessed must be objective. Subjective evaluations because of a personal preference for any time can be extremely hurtful, erroneous, and moreover unethical.

In general, the professionals who are biased toward theory tend to produce data that are often abstract, and the intellectual contribution is expressed in the form of theories, with or without proof. Other than an equally well-versed peer, when proof is absent, there is typically no one at the management level who can evaluate such work. As a result, publication in the proceedings of a conference may be the only outlet for such efforts, after which publication in a *reputable* journal may be possible but only with considerable effort or, for various reasons, may not be possible at all.

Publication of data in the proceedings from a conference often results in a shorter time to release of the data and a shorter time print (of the conference proceedings). This follows from the opportunity to describe competed or partly completed work before peer scientist and/or engineers and to receive a more complete review than the type of review that is typical for a journal. At a conference, the audience asks general and specific questions of the presenter that often guide the presenter into further work or to diverge into a new line of investigation. Overall, this will help the presenter to finalize the document for publication in the proceedings (where the proceedings are published postconference).

On the other hand, some scientists and engineers wonder if journal reviewers really pay attention to the salient points of the potential publication or if they merely look for errors in style and grammar, which often are of some importance but have no bearing on the technical quality of the submitted manuscript. Several readers may have experienced all of the above behavior on the part of reviewers. However in many academic reviews, statements are made that publication in the proceedings of a prestige conference is inferior to publication in a prestige journals, without the writers realizing or even being willing to admit that in relation to data presentation and publication, many conferences are superior to an established journal.

There is also the scientist and engineer who wishes to publish his/her work and who may feel stymied because of (1) a company policy related to proprietary material—a justified reason, or (2) an arbitrary uninformed and incorrect (sometimes personal) decision by a supervisor or a member of the company review committee—an unjustified reason. However, for the nonacademic scientist and engineer, there is the medium of publication of the material as a company report. This can be a worthwhile method for circulating one's work throughout the company, providing that company management can recognize the value of the work and that the manager has not yet reached his/her highest level of incompetence (Peter and Hull, 1969; Lazear, 2000; Pluchino et al., 2010)—often called the *Peter Principle* (Chapter 7). In addition, the importance of the work to the young scientist and engineer can, again, be diminished and the names of a supervisor and any other persons higher up the food chain are included as coauthors.

The educated young professional scientist and engineer may often wonder if s/he is merely a pair of hands (for an overbearing and/or incompetent supervisor, or an overbearing and/or incompetent department head, or jealous and/or incompetent colleagues, and/or all of the prior examples) and is not supposed to be given credit for the ability to think and solve a problem. Performance suffers and, with repeated negativism toward publication, the young professional starts to lose interest in the organization. Lack of recognition for hard and intelligent work is a killer and getting the best out of any such scientists and engineers becomes an impossible dream.

In any employment situation, the scientist and/or engineer may be required to establish that s/he has performed better insofar as the work can be established as being *better* than that of any predecessor and has new or novel aspects that are of value (to the company). Put simply, "better" can mean (1) a process for producing an improved product, (2) completion of a project under budget, or (3) anything in between these two limits.

"Better" does not mean, by any stretch of the imagination, basing the outcome of the evaluation on the ability of the members of the evaluation committee or the supervisor to count the number of publications—without using the fingers! However, if "better" is defined or translated in this manner, the fundamental basis for deciding on achievement may be lost because of a complete misunderstanding of the nature of the work and its importance to the university or to the company. The full effect of the contribution is lost in the argument of word definitions. As a result, the young scientist or engineer sees that his/her work is being recognized as meaningful. Morale improves and the young professional is satisfied, for the moment! However, the means by which the impact of the contribution can be assessed is, in fact, a major issue.

Most of all, the evaluation process should involve knowledge of the education of scientist and engineers as well as being able to speak their language. However, there are other aspects to getting the best out of scientists and engineers and these relate to the evaluation process.

In academia, the young professional enters a department at the assistant professor grade. At this level, the assistant professor has little choice in terms of choosing teaching assignments and has administrative work thrust upon his/her shoulders— and can complain about the workload at his/her peril—while the older tenured members of staff have the right to refuse such work without fear of any form of reprisal.

In addition, the assistant professor also has to acquire research funding and may even have to pass his/her reports/papers through a review committee prior to publication. This review committee will be made up of senior members of staff who, for many reasons that are often difficult to follow, can give the young professor a glowing performance report or a report that is somewhat less than glowing. In many cases, the report may be subjective rather than objective. The latter is of some use to the assistant professor—it is as a result of a subjective assessment that the young assistant Professor can feel that s/he is suffering rejection by his/her colleagues and s/he becomes determined to move at the first/best opportunity. On the other hand, an objective assessment can be very helpful and show the assistant professor that his/her work is recognized and appreciated or that there may be holes in the work that should be filled before finalization of the manuscript.

In addition to the scientist or engineer who may not appear to fit the academic or company mold, the evaluation process may seem to focus on the *do-as-I-say* dictum of the immediate supervisor, department head, or academic senior colleagues. In academia, the additional dictum of *publish-or-perish* is also operative, insofar as *publish* has the standard academic meaning of *publish in recognized journals*.

In the first instance (i.e., the *do-as-I-say* syndrome), the young professional may find that s/he is up against a brick wall. The supervisor/department head is all powerful and the scientist or engineer has little or no recourse for appeal. Equally, the *do-as-I-say* syndrome is also fraught with pot holes for the young scientist and engineer. Production of patentable work requires acknowledgment of the supervisor and any other designee as coauthors, is also ruinous to the young scientist and engineer. But where are these worthies if the work does not product patentable ideas and the project is terminated. Where is blame assigned? To the young scientist and engineer! The supervisor and other potential designees have backed away and are not evident by any form of presence or support.

Counting the number of publications in recognized journals ignores the quality of any particular publication as well as the potential for benefit through ownership (by the university) of intellectual property. For example, publication of work in a patent followed by publication of the work in a conference proceedings are tangible means of conveying ideas and insight that relate to intellectual property. Obligating scientists and engineers to be evaluated without giving true credence to intellectual property is a handicap and is ruinous of a true method of evaluation.

Neither of these scenarios is a way to encourage either academic achievement or industrial achievement in science and engineering. The scientist and engineer must not be given cause to wonders if s/he is merely a pair of hands for a supervisor and must be given credit for the ability to think and solve a problem. Lack of recognition for hard and intelligent work is an unprofessional attitude on the part of any supervisor/mentor and getting the best out of any such scientist or engineer becomes an impossible dream. The young scientist or engineer sees that his/her work is being recognized as meaningful. Morale improves and the young professional is satisfied, for the moment! The work must be assessed fairly and the significance and impact of the work must be recognized.

Counting the number of publications in recognized journals as the main means of evaluation of the work of a scientist and engineer may cause the evaluation committee to ignore the quality of any particular publication as well as the potential for benefit through ownership (by the university) of intellectual property. Obligating scientists and engineers to be evaluated without giving true credence to intellectual property is a handicap and either directly is ruinous of a true method of evaluation. Without recognition, performance of scientists and engineers will suffer, morale will deteriorate, and the young professional starts to lose interest in the organization. This can be a serious blow to the morale of scientists and engineers because some of the field's best researchers may work at other institutions.

### 1.5.2 RECOGNITION

The scientific and engineering fields are composed of educated and relatively young professionals who have the ability to apply themselves to the problems at hand, whether theory studies or experimentation. To the scientist and engineer, the outcome of this work that offers some form of gratification is (1) completion of a project and (2) publication of the data in a journal or similar medium for distribution to one's peers. The latter gives the scientist and engineer recognition for their work.

In general, the scientific and engineering professionals who are biased toward theory tend to produce data that are often abstract and the intellectual contribution is expressed in the form of theories with proof. As a result, publication on the proceedings of a conference may be the only outlet for their efforts, after which publication in a reputable journal may be possible but only with considerable efforts or, for various reasons, may not be possible at all. For the nonacademic scientist and engineer, there is the medium of publication of the material as a company report. This can be a worthwhile method for circulating one's work throughout the company. But, the importance of the work to the young scientist or engineer can, again, be diminished if/when the names of a supervisor and any other persons higher up the management chain are included as coauthors, without good or just cause.

Publication of data in the proceedings from a conference often results in a shorter time to print. This follows from the opportunity to describe completed or partly completed work before peer scientists and/or engineers and to receive a more complete review than the type of review that is typical for a journal. At a conference, the audience asks general and specific questions to the presenter that often provides recommendations for further work or a new line of investigation. Overall, this will help the presenter to finalize the document for publication in the proceedings (where the proceedings are published postconference).

On the other hand, in academia, the young professional enters a department at the assistant professor grade. At this level, the assistant professor has little choice in terms of choosing teaching assignments and has administrative work thrust upon his/her shoulders while the older tenured members of staff have the right to refuse such work without fear of reprisal. In addition, the young Assistant Professor also has to acquire research funding and may even have to pass his/her reports/papers through a review committee prior to publication. This review committee will be made up of senior members of staff who, for many reasons that are often difficult to follow, can give the young professor a glowing performance report or a report that is somewhat less than glowing. It is at this time, if the latter is the case, that the young professor can feel that s/he is suffering rejection by their colleagues.

The educated young professional scientist and engineer wonders if s/he is merely a pair of hands (for an overbearing supervisor or an overbearing department head) or the focus of malicious behavior on the part of jealous colleagues and is not supposed to be given credit for the ability to think and solve a problem, without input from any of the aforementioned persons. Performance suffers and, with repeated negativism toward publication, the young professional starts to lose interest in the organization. Lack of recognition for hard and intelligent work is a deterrent to excellence in work performance and getting the best out of any such scientists and engineers becomes an impossible dream.

The principle of fairness and the role of personal recognition within the reward system of science account for the emphasis given to the proper allocation of credit (Chapter 3). In the typical scientific or engineering paper, credit is explicitly acknowledged in three places: (1) in the list of authors; (2) in the acknowledgments of contributions from others, usually minor contributions and many little more than talking out loud discussions or acting as a sounding post; and (3) in the list of references or citations. Conflicts over proper attribution can (and often do) arise in any of these categories.

Citations serve many purposes in a scientific or engineering paper—(1) they acknowledge the work of other scientists and engineers, (2) direct the reader toward additional sources of information, (3) acknowledge conflicts with other results, and (4) provide support for the views expressed in the paper. More broadly, citations place a paper within its scientific context, relating it to the present state of scientific knowledge (NAS, 1995). Failure to cite the work of others can give rise to more than just hard feelings. Citations are part of the motivation (Chapter 6) and reward system of science and engineering. They are connected to funding decisions and to the future careers of researchers. More generally, the misallocation of credit undermines the incentive system for publication.

In addition, scientists and engineers who routinely fail to cite the work of others may find themselves excluded from the fellowship (technical meetings) of their peers, or even lunch-time and/or break-time conversations. This consideration is particularly important in one of the more intangible aspects of a scientific or engineering career–that of building a reputation. Published papers document the approach that a scientist or engineer has applied to the project/problem, which is why it is important that they be clear, verifiable, and honest. In addition, a scientist or engineer who is open, helpful, and full of ideas becomes known to colleagues and will benefit much more than someone who is secretive or uncooperative (Speight and Foote, 2011).

### 1.5.3 Teacher/Professor Evaluation

The first step of the evaluation should be to determine if the teacher/professor has accreditation in the subject that s/he is teaching (Chapter 4) (TEAC, 2014). In other words, there should be a serious effort to determine if s/he is qualified (certified) to teach that subject. If not, there must be serious action taken to either ensure that the teacher/professor moves to get accreditation or that s/he should be removed from that classroom. Once this issue has been addressed, then there is the move to the evaluation proper. However, if the teacher/professor does not have accreditation in the subject under discussion, steps must be taken to add relevant comments to the evaluation of the school principal/department head/dean for allowing an unaccredited teacher/professor the teaching/professorial freedom of the classroom.

While students must expect to be subject to periodical evaluation through the mechanism of closed-book examinations, assignments, classroom performance, and teacher/professor–student contact, there must also be in place and active a mechanics for teacher/professor evaluation. It is appropriate to include such a discussion at this point in the text since it does influence the outcome of many of the situations described in the following chapter.

Many universities (and some high schools) request (insist) that each student completes a course evaluation in writing (and presented as a *course questionnaire*) which generally focuses on course content, usefulness to the student, and also evaluation of the abilities and subject knowledge of the teacher/professor. These are typically competed anonymously and submitted at the end of the course—students can be very distrustful of the word "anonymous." In theory, such evaluations are meant to offer guidance to the teacher/professor for any omissions in the course and to the administrators to indicate the quality of the teacher/professor. Hopefully the teacher/professor is not a biased person if the names of the students become known.

After all, any student must be regarded as a paying client (or his/her parents are the paying clients either directly through fees to the institution or through the structure of [federal or state or county] taxes) and has a right to receive quality teaching for the money paid. Many teachers do not recognize this aspect of student–institute relationship and treat the students as lower class citizens who have the effrontery to encroach on their day by expecting work, performance, and teaching excellence. Of course, and as a result of attitude issues, the evaluations of the teacher/professor by the students are (not always but) often (1) ignored, (2) lost on the bus, (3) stolen from the car, (4) read and discarded, (5) thrown into the nearest garbage can, (6) used to light the campfire on a weekend trip to the great outdoors, (7) used as a Kleenex tissue, or (8) all of the above.

Hence credibility for the evaluation system as it pertains to the performance of teachers and professors is extremely low. Yet, credibility in any evaluation system is essential. A school principal/school superintendent/university president/dean/department head must be able to report to the powers that be that s/he knows of the quality of the teachers/professors by virtue of the quality and credibility of the evaluation system (Stensaker, 2013). For example, to assess the quality of teachers/professors it is very necessary to *define* quality teaching. It is not sufficient to acknowledge the years of service of a teacher/professor and conclude that because of the large number of years of service his/her teaching must be good or that the ability of the teacher/professor to arrange funding sources has anything to do with good teaching. The system used for evaluation must include not only the teaching that occurs in the classroom but also the behind-the-scenes work of planning and other professional work, such as communicating with families or students and participating in a professional community. For each component of the evaluation system there should be (at least) four levels of performance: (1) unsatisfactory, (2) basic, (3) proficient, and (4) distinguished, which describe the degrees of teacher/professor expertise in that component.

There must also be a shared understanding of the evaluation process. *Everyone* in the system—teachers, professors, mentors, supervisors, department heads, deans, and even the university vice presidents and president—must possess a shared understanding of this system. For the president of a university to cavalierly state that it is not his/her prerogative to particulate in the evaluation system is to shirk his/her duty. Having a common language, from the top person down or from the bottom person up, to describe the evaluation practice increases the value of the conversations that ensue from classroom observations. As an example, discussing student engagement in learning is more effective when everyone understands the meaning of such a catchword phrase in light of four elements: (1) activities and assignments, (2) grouping of students, (3) instructional materials and resources, and (4) structure and pacing. Conversations using this more specific language invite teachers/professors to critically analyze their own practice and invite observers to inquire about the decisions a teacher/professor has made in planning and executing a lesson.

Those persons who perform evaluations—particularly those higher up the administrative or experience ladder—must be able to recognize classroom examples of the different components of practice, interpret that evidence against specific levels of performance, and engage teachers/professors in productive conversations about their teaching practice. Evaluators must be able to assess teachers/professors accurately so that they (teachers/professors) accept the judgments as valid and the public has confidence in the results.

Evaluations that focus on quality assurance (unless there is extreme positive or negative bias on the part of the evaluator) typically yield judgments that are fair, reliable, valid, and (above all) acceptable. They are helpful in determining whether the skill of a teacher/professor has slipped below standard and needs strengthening. School and university administrators may then use the evaluations for decisions regarding continuation of employment or termination of employment and compensation. This is crucial when deciding (1) which teachers/professors should attain permanent status as tenured professionals, (2) which teachers/professors should have their employment terminated, and (3) which teachers/professors should be nominated for leadership positions as mentors or coaches. After all, is this not the way of the real world when scientists and engineers work in commercial employment?

The key, of course, is to institute a credible system of teacher evaluation that requires higher levels of proficiency of evaluators than the old checklist observation model. Evaluators need to be able to assess accurately, without bias, and provide meaningful feedback, as well as engage teachers in productive conversations about practice. As worthy a goal as this may seem, there may be howls of protest from a number (not all) of the teachers/professors, often without good reason, other than they are being placed in a position of accepting responsibility for their teaching activities and output in the form of student advancement.

A thoughtful approach to teacher/professor evaluation—one that engages teachers in reflection and self-assessment—can yield benefits far beyond the important goal of quality assurance. Such an approach provides the vehicle for teacher/professor growth and development by providing opportunities for professional conversation around agreed-on standards of practice. The same principles should be applied to the evaluation of teaching assistants—typically, a teaching assistant (TA) is a *graduate teaching assistant* (GTA) who is a registered full-time graduate student chosen as a result of excellent scholarship and promise as a teacher. In the context of this book, the teaching assistant should hold a baccalaureate degree in the scientific or engineering field that is closely related to the one in which s/he will be assisting. Under faculty supervision, the duties of a teaching assistant may include (1) providing help sessions and leading discussion, recitation, laboratory, or quiz sections; (2) holding office conferences with students; (3) preparing materials for faculty-guided classroom or laboratory instruction; (4) assisting the professor to the design of a course; (5) assisting in the design and/or preparation of exams or quizzes; (6) and proctoring examinations and grading student papers and/or examinations to faculty-guided standards (Chapter 2) (Lyall, 1997).

While there are excellent standards for evaluating teaching assistants (Lyall, 1997), evaluations do not always occur on any basis, let alone on a regular basis, and the ideas of the students must also be taken into account (Hazari et al., 2003; Hazari and Key, 2006). Typically, the teaching assistants are given a free hand and allowed to teach the students in the manner that they know best, which is often ineffective. Some teaching assistants do have the skills necessary to teach a course. Other teaching assistants are not always knowledgeable in the subject matter of the course and may be only one lesson ahead of the students! And since the students are paying fees for the service and teaching by a professor, appointing a teaching assistant (especially an inexperienced teaching assistant) may not fulfill the agreement between the student and the university in which the student expected a better level of teaching.

## 1.6  THE IMPACT OF SCIENCE AND ENGINEERING

Though modern science and engineering are of relatively recent origin, having started with the Renaissance of the fifteenth century, both disciplines have made rapid progress through the last four centuries and have completely transformed the manner of technological progress. Some observers note (with justification) that technology has changed more in the last 400 years than at any time in the millennia since the time of the Sumerians (the common name given to the ancient non-Semitic inhabitants of Southern Mesopotamia, also known by the Biblical term *the Plains of Shinar*, currently in southern Iraq) who lived, worked, and studied in the approximate period 3000–4000 BC. The application of scientific and engineering knowledge accumulated over the last 400 years has had a high and very visible impact on the advancement of technology.

In spite of this progress, and of the consequent development of technology and industry, and of the conveniences, comforts, and power that have been acquired through scientific and engineering knowledge, peace is not a worldwide commodity and many human beings are unable to live without the presence of continued violence. In fact, many scientific and engineering discoveries have been applied to the advancement of military technologies. Furthermore, during the Victorian age, it was hoped that the development and spread of science and engineering would usher in an era of peace and prosperity, but that has not been the case. On the contrary, the slaughter that occurred during the U.S. Civil War and also during World War I can be ascribed predominantly to the application of science and engineering know-how to the development of military technology, such as rapid-firing guns (the Gatling gun of the late Civil War period and the machine gun of World War I), more devastating explosives, and poison gas (Tuchman, 1962; Freemantle, 2012). The military incompetence of the commanding generals must not be forgotten as they failed to recognize the advance in military technology and caused the unnecessary deaths of hundreds of thousands of soldiers—the military command structure of the top brass consisted of generals who were living in the past and they must also share some of the responsibility for the slaughter.

The greater the level of violence (through wars and increasing crime) throughout the world during the decades since the end of World War II, the greater the level of prosperity—so-called *globalization* (Chubin et al., 2005). Moreover, there has been continued violence and intergovernmental tensions (e.g., the Cold War, which ran from 1945 to 1980, give or take a year or two), and newer diseases as well as the spread of older little-known diseases.

In spite of the subjects of the above paragraphs, the object of science and engineering is to extend knowledge of the physical, biological, or social world beyond what is already known—and this has been a successful trait throughout the nineteenth century and the twentieth century. But the knowledge gained by an individual through experience can only properly enter the domain of science or engineering after it is presented to others in such a fashion that they can independently judge its validity. This process occurs in many different ways insofar as scientists and engineers—as far as nondisclosure agreements will allow—can (1) talk to their colleagues and supervisors in laboratories, in hallways, around the proverbial water cooler, and over the telephone; (2) deliver presentations at seminars and conferences; (3) and publish results in scientific journals and engineering journals, which in turn send the papers to be scrutinized by reviewers. After a paper is published or a finding is presented, it is judged by other scientists in the context of what they already know from other independent sources. Throughout this continuum of discussion and deliberation, the ideas of individuals are collectively judged, sorted, and selectively incorporated into the consensual but ever evolving scientific worldview. In the process, individual knowledge is gradually converted into generally accepted knowledge and thence into practice.

Thus, well-qualified professional scientists and engineers with advanced degrees (Chapter 2) play a central and growing role in industrial and commercial life. They contribute directly to the national goals of technological, economic, and cultural development, not only as researchers and educators but in a wide variety of other professional roles. However, there are critical issues relating to the assessment of impact of the work of scientific and engineering professionals: (1) conference proceedings aside, the impact of any new technology or modified technology takes time; and (2) a measure of the impact is not achieved by the use of a so-called standard citation index. There is always the distinct possibility that the number of citations is directly related to those who are critics of the work and may consider it nothing short of ludicrous if it is apparent that the reviewers concentrated on grammatical errors rather than on scientific content. Not all papers in high-quality publications are of great significance, and high-quality papers can appear in lower quality publication media. Therefore, the academic form of evaluation—which generally considers the number of publication (often irrespective of quality) as a major determinant for promotion—can be severely underwhelming and incorrect (Shola Omotola, 2013).

On the other hand, the supervisor of the young professional may fail (or refuse) to recognize the impact of the work, especially if his/her name (i.e., the name of the supervisor) is not included as a coauthor. The answer to this dilemma is difficult and must remain in the dark recesses of the mind of many supervisors/mentors.

The scientist or engineer may request that to evaluate his/her contribution and its effects, evaluators should be selected from academic or company institutions if the concept of objective evaluation has been reduced to practice. Some academic institutions and companies prohibit such methods of evaluation from writers not having an academic affiliation or a company affiliation. In the case of a company-requested evaluation, the company runs the risk of an evaluator from outside of the company becoming privy to otherwise protected intellectual property; the evaluator must therefore be from within the company, giving the evaluation a taint of being subjective and somewhat incestuous. This can be a serious blow to the morale of scientists and engineers because some of the field's best researchers work at other institutions.

Unless such an objective assessment of the work can be performed, the young scientist or engineer is inclined to believe that the significance of his/her work may be ignored. Thus, in the minds of the academic or company peers, the significance and impact of the work is not recognized and work that embodies new ideas or techniques is bypassed. A fair assessment of the work of any young scientist or engineer is not always possible!

## REFERENCES

AAUP (American Association of University Professors). 2006. 1940 statement of principles on academic freedom and tenure with 1970 interpretive comments. In: *AAUP Policy*, 10th edn. American Association of University Professors (AAUP), Washington, DC.

Andreescu, L. 2009. Foundations of academic freedom: Making new sense of some aging arguments. *Studies in Philosophy and Education*, 28(6): 499–515.

Batterbury, S.P.J. 2008. Tenure or permanent contracts in North American higher education? A critical assessment. *Policy Futures in Education*, 6(3): 286–297.

Boyer, E.L. 1990. *Scholarship Reconsidered: Priorities of the Professoriate*. The Carnegie Foundation for the Advancement of Teaching, Princeton, NJ.

Brown, J. 1992. *The Definition of a Profession: The Authority of Metaphor in the History of Intelligence Testing, 1890–1930*. Princeton University Press, Princeton, NJ.

Chubin, D.E., May, G.S., and Babco, E.L. 2005. Diversifying the engineering workforce. *Journal of Engineering Education*, 34: 73–86.

Conrad, R. 1993. *Academic Freedom*. Routledge Publishers, Taylor & Francis Group, New York.

Derrida, J. 1988. *Limited Inc*. Northwestern University Press, Evanston, IL.

ECPD (Engineers' Council for Professional Development). 1941. The Engineers' Council for Professional Development. *Science*, 94(2446): 456.

Freemantle, M. 2012. *Gas! Gas! Quick, Boys!: How Chemistry Changed the First World War*. The History Press, Charleston, SC.

Fullmer, J.Z. 1969. *Sir Humphry Davy's Published Works*. Harvard University Press, Cambridge, MA.

Fullmer, J.Z. 2000. Young Humphry Davy: The making of an experimental chemist. *American Philosophical Society*, 237: 158.

Garrett, R. 2013. What is teacher tenure? http://www.education.com/magazine/article/what-is-teacher-tenure/(accessed August 26, 2013); http://www.education.com/debate/teacher-tenure/?results# (accessed August 26, 2013).

Glencorse, B. 2013. Testing new tools for accountability in higher education. In: *Global Corruption Report: Education* (Chapter 4.14). Routledge, Taylor & Francis Group, Abingdon, Oxford.

Grau, C. 1988. *Berühmte Wissenschaftsakademien: Von ihrer Entstehung und ihrem weltweiten Erfolg (Famous Academies of Science: From Its Emergence and Its Worldwide Success)*. H. Deutsch Gmbh, Frankfurt am Main, Germany.

Hamilton, J. 2004. *A Life of Discovery: Michael Faraday, Giant of the Scientific Revolution*. Random, House, New York.

Hazari, Z., Key, A.W., and Pitre, J. 2003. Interactive and affective behaviors of teaching assistants in a first year physics laboratory. *Electronic Journal of Science Education*, 7(3): 1–38. http://wolfweb.unr.edu/homepage/crowther/ejse/hazarietal.pdf (accessed August 6, 2013).

Hazari, Z. and Key, T. 2006. Student evaluations of teaching assistants in a first year physics laboratory: Is there a grade dependence? *Physics in Canada*, H1–H4. http://www.clemson.edu/ese/per/wp-content/themes/gallery/gallery/images/publication7.pdf (accessed August 6, 2013).

Hedrick, D.R. 2009. *Technology: A World History*. Oxford University Press, Oxford.

Hill, N. 2014. *Pressure to Perform*, pp. A1–A3. Laramie Daily Boomerang, Laramie, WY.

Hofstadter, R. and Metzger, W.P. 1952. *The Development of Academic Freedom in the United States*. Columbia University Press, New York.

Holcomb, B., Kay, J., Kay, P., and Monk, J.J. 1987. The tenure review process. *Journal of Geography in Higher Education*, II(2): 85–98.

Lammas, J.C. and Garcia, M.A. 2009. Exploring the concept of "Profession" for organizational communication research institutional influences in a veterinary organization. *Management Communication Quarterly*, 22(3): 3357–3384.

Landels, J.G. 1978. *Engineering in the Ancient World*. University of California Press, Berkeley, CA.

Lazear, E.P. 2000. *The Peter Principle: Promotions and Declining Productivity*. Revision 10/12/00. Hoover Institution and Graduate School of Business, Stanford University, Stanford, CA. http://www-siepr.stanford.edu/Papers/pdf/00-04.pdf (accessed July 22, 2013).

Lyall, K. 1997. *Teaching Assistant Evaluation and Improvement Handbook*. University of Wisconsin System Board of Regents, Engineering Learning Center, College of Engineering, Engineering Hall, University of Wisconsin, Madison, WI. http://www.theiet.org/membership/profreg/ieng/ (accessed August 5, 2013).

Martin, M. and Schinzinger, R. 2005. *Ethics in Engineering*, 4th edn. McGraw-Hill, New York.

McKenzie, R.B. 1996. In defense of academic tenure. *Journal of Institutional and Theoretical Economics*, 152(2): 325–341.

Michaels, P.J. 2004. *Meltdown: The Predictable Distortion of Global Warming by Scientists, Politicians, and the Media* (Chapter 11). The Cato Institute, Washington, DC.

NAS (National Academy of Sciences). 1995. *On Being a Scientist: Responsible Conduct in Research*, 2nd edn. National Academy of Sciences, National Academy of Engineering, Institute of Medicine, Washington, DC.

NAS (National Academy of Sciences). 1997. *Advisor, Teacher, Role Model, Friend: On Being a Mentor to Students in Science and Engineering*. National Academy of Sciences, National Academy of Engineering, Institute of Medicine, Washington, DC.

Olgiati, V. 2010. The concept of profession today: A disquieting misnomer? *Comparative Sociology*, 9(6): 804–842.

Pasachoff, N., and Pasachoff, J.M. 2012. Galileo Galilei: Laying the foundations of modern science. In: A. Robinson (ed.), *The Scientists: An Epic of Discovery*. Thames & Hudson, London.

Peter, L.J. and Hull, R.R. 1969. *The Peter Principle: Why Things Always Go Wrong*. William Morrow and Company, New York.

Pluchino, A., Rapisarda, A., and Garofalo, C. 2010. The Peter principle revisited: A computational study. *Physics A: Statistical Mechanics and its Applications*, 389(3): 467–472.

Riley, N.S. 2011. *The Faculty Lounges and Other Reasons Why You Won't Get the College Education You Paid for*. Ivan R Dee, The Rowman Littlefield Publishing Group, Lanham, MD.

Robinson, A. (ed.). 2012. *The Scientists: An Epic of Discovery*. Thames & Hudson, London.

Robinson, D. 2013. Corrupting research integrity. In: *Global Corruption Report: Education* (Chapter 3.18). Routledge, Taylor & Francis Group, Abingdon, Oxford.

Russell, C.A. 2000. Michael Faraday: Physics and faith. In: O. Gingerich (ed.), *Oxford Portraits in Science Series*. Oxford University Press, Oxford.

Salmi, J. and Helms, R.M. 2013. Governance instruments to combat corruption in higher education. In: *Global Corruption Report: Education* (Chapter 3.2). Routledge, Taylor & Francis Group, Abingdon, Oxford.

Searle, J.R. 1971. *The Campus War: A Sympathetic Look at the University in Agony*. World Publishing, London.

Shallcross, D. 2005. Chemical engineering education. In: *Chemical Engineering and Chemical Process Technology*. Encyclopedia of Life Support Systems (EOLSS), Developed under the Auspices of the UNESCO, EOLSS Publishers, Oxford.

Shaw, M.M. and Despota, K. 2013. Journals: At the front line of integrity in academic research. In: *Global Corruption Report: Education* (Chapter 3.20). Routledge, Taylor & Francis Group, Abingdon, Oxford.

Shola Omotola, J. 2013. Corruption in the academic career. In: *Global Corruption Report: Education* (Chapter 3.15). Routledge, Taylor & Francis Group, Abingdon, Oxford.

Singer, C., Holmyard, E.J., Hall, A.R., and Williams, T.I. (eds). 1954. *A History of Technology*. Oxford University Press, Oxford.

Smallwood, S. 2003. Tenure Madness. *The Chronicle of Higher Education*, 49(37): A10.

Speight, J.G. 2011. *The Refinery of the Future*. Gulf Professional Publishing, Elsevier, Oxford.

Speight, J.G. and Foote, R. 2011. *Ethics in Science and Engineering*. Scrivener Publishing, Salem, MA.

Speight, J.G. 2014. *The Chemistry and Technology of Petroleum*, 5th edn. CRC Press, Taylor & Francis Group, Boca Raton, FL.

Stensaker, B. 2013. Ensuring quality in quality assurance. In: *Global Corruption Report: Education* (Chapter 3.5). Routledge, Taylor & Francis Group, Abingdon, Oxford.

Sykes, C.J. 1988. *ProfScam: Professors and the Demise of Higher Education*. Regnery Gateway Publishers, Washington, DC.

TEAC (Teacher Education Accreditation Council). 2014. http://www.teac.org/ (accessed June 1, 2014).

Thomas, J.M. 1991. *Michael Faraday and the Royal Institution: The Genius of Man and Place*. Hilger Publishers, Bristol, United Kingdom.

Tuchman, B.W. 1962. *The Guns of August*. The Macmillan Publishing Company, New York.

Wright, M.T. 2005. Epicyclic gearing and the Antikythera mechanism, Part 2. *Antiquarian Horology*, 29(September 2005): 54–60.

# 2 Education of Scientists and Engineers

## 2.1 INTRODUCTION

Scientists and engineers have specialized skills that are applied to resolving problems related to basic research, applied research, and teaching, from which knowledge learned is then applied to the development of new technologies. Nonscientists and nonengineers are not omitted from this scenario as they may—although they may not be actively employed in research or product development—bring an understanding of science and engineering to other occupations and professions.

The key issue in any such situation is education, which is the fundamental basis that eventually leads to achievement of state goals (in the United States) and national goals (not only in the United States but also in many other countries). But first, schools and universities must be held responsible for producing qualified teachers and research personnel (professional and technical), but this is not always the case and improvements are needed (Curtis, 2014). By producing teachers and educating students in the context of research, the systems for the education of scientists and engineers prepare them for research careers in academia, government, and industry. Furthermore, by attracting outstanding students and faculty members (hopefully who have some understanding of the nonacademic world) a national education system can (hopefully) show benefit as a result of an influx of talent as well as an influx of innovative ideas. In fact, an increase in scientific and engineering knowledge and the manner in which this knowledge is applied is essential for successfully developing new technologies and industries, developing new sources of energy, and reducing environmental pollution (Speight and Foote, 2011).

Furthermore, the future of the scientific and engineering workforce begins with individual decisions by the institutions of learning to select and prepare students to make the necessary decisions that lead to the choice of a career in science or engineering. Among the factors that are important to this decision are: (1) the type or types of school attended and courses taken, (2) the teaching practices employed, (3) the ability and credibility of the teachers/professors, (4) the ability and talent of the student, (5) the type of undergraduate institution attended, (6) early participation in scientific or engineering research, (7) and the availability of funding. In some cases, however, gender, race or ethnicity, parental occupations, and socioeconomic status can also play decisive roles (OTA, 1988a; Leach, 2013).

For example, the result of recent surveys shows that, in the life sciences and physical sciences, almost 56% of doctorates were awarded to male candidates, while 78% of the candidates receiving a doctorate in engineering from universities in the United States were male. This has shown a small but steady improvement each year for the past 30 years but still leaves an obvious gender bias in engineering (NRC, 1994; NSF, 2012a,b; Leach, 2013). In the same time period, the number of students receiving master's degrees from universities in the United States, including those who intended to continue toward a doctorate, has also increased steadily over the past 30 years (NSF, 2012a,b). However, a substantial number of the recent increase in the total science and engineering doctorates awarded annually can be accounted for by an influx of foreign students. Because of this influx, the average growth in the total science and engineering graduate-student population—including foreign students—has shown an approximate 2.5% increase per year since 1982 (NSF, 2012a,b; Fiegener and Proudfoot, 2013).

In reality, the greatest influence on the size and quality of the science and engineering work force is elementary education and high-school education (Curtis, 2014). Because of relatively poor teacher training, schools and institutions of higher learning must shoulder much of the blame for failing to instill interest in science and engineering. Schools—through the employment of qualified teachers—must work to generate interest in science and engineering and prepare students with the necessary background in science and mathematics (OTA, 1988b). The remainder of the blame must fall squarely on the shoulders of the institutes of higher education (the universities) and the poor quality of teaching therein, because of the failure to provide the necessary training programs and to mentor the students correctly and meaningfully (Chapters 4 and 6). Unfortunately, this tier-type education system sets up a situation where an institute of higher education has a ready-made excuse for poor performance in teaching (and mentoring) by loudly and unashamedly expressing disappointment about the quality of students emerging from the schools into the university system. The converse is also true, in which the schools can blame the universities for failing to educate students.

Thus, each part of the education process needs to recognize that the task is not only to do the best with the available students rather than bemoan the situation and lay blame everywhere but where blame should be laid, but also to provide auditable teacher training. The old adage "it is not my fault, it is his/her fault" works well within the education system, especially when the excuse is hidden in a myriad of meaningless words and sentences! There are very few references to anyone in academia who has ever admitted that it is their fault. It may well be that the university entrance system is at fault for the poor quality of higher education—accepting students without much (any) effort being put into a selection process (some universities, but not all, do have a commendable selection process) because "we need the money" or "we want the money"—either adage is not an acceptable excuse for the poor quality of higher education but it is a reason for the poor quality of higher education. And to make matters even worse, some of these students of less-than-acceptable quality, who seem to be able to pass exams (without the marks being made public) or performance being monitored, may even move on to become teachers or professors!

The training of science and engineering teachers—professors are also included in this group—with the common emphasis on teaching methods is often ignored. Often, the methods of the delivery of knowledge in the classroom (at school or university) fail to impress upon the minds of future science and engineering teachers or researchers an understanding of and enthusiasm for science and engineering as being processes of inquiry. Instead, the subject matter comes across to the students as a collection of uninteresting and unintelligible facts which—as perceived by the students—seem to bear no relationship to reality and certainly do not offer enticing prospects for a career (Matthews, 2007).

## 2.2 THE PREUNIVERSITY EXPERIENCE

In the preuniversity system (i.e., the school system), a teacher or school teacher is a person who provides educational instruction for pupils (children) and students (young adults or adults). In addition, it is to be expected that a teacher will use a lesson plan to facilitate student learning in providing a course of study (curriculum). In many countries, a person who wishes to become a teacher must first obtain specified professional qualifications (a degree) from a recognized university and/or teacher training college, which is often college within a university, and the professional qualifications may include the study of pedagogy—the science and art of education, which ultimately leads to the acquisition of skills. Furthermore, teachers, like any other group of other professionals, may (should) have to continue their education after they qualify (*continuing professional development*).

Just as universities and other institutes of higher education must prepare students to become the next generation of scientists and engineers, the teaching atmosphere (and mentoring) at any level of the school education system will have an influence on the thinking that goes into the choice of a career by a student (OTA, 1989). Indeed, it is not always easy to identify the multitude of factors that encourage a student to choose a scientific and engineering career. It is often equally difficult to enumerate the factors that deter a student from choosing a career in science or engineering. In fact, many scientists and engineers develop a career interest as early as elementary or junior high school and many may have made definitive choices even before entry into the high-school system. This has led to the indication that some future scientists and engineers select these majors early in life, then work hard and persist with their career plans without considering any other options. On the other hand, some students, after choosing their major subjects early in school life, make a change during high school (OTA, 1988a). Only careful guidance and mentoring can assist the young student to find the pathway to education that will satisfy his/her career goals.

Universities rely on elementary and secondary schools to produce a cadre of students with the potential to be successful on the career track and to graduate with the degree of choice to join that portion of the workforce composed of professional scientists and engineers. If there is insufficient preparation in the preuniversity schools and development of an interest in science and particularly in mathematics wanes, students will not be able to succeed in science or engineering in a university. Assuming that the school can provide an environment for developing the ability of students, no single school environment is suitable for all students. Some school administrators may feel that the school produces more potential scientists and engineers than others, but the reality is whether or not these students continue along the track to graduate in science or engineering. What the teacher sees and what the capabilities of the student may be are often two distinctly different items. This is especially true if the teacher is unqualified or has not been certified to teach science or engineering subjects. But this is the very sharp two-edged sword that results from the tier-type education system. Having done their bit, the schools can now blame the university failing to develop students to their fullest potential, as defined by the school(s). Pity the poor students who have not received any meaningful mentoring or guidance and are destined to fail because of the shortcomings of the school–university exchanges.

The future supply (and quality) of scientific and engineering graduates depends on how well the teaching staff at preuniversity schools encourage students (but not against the will of the students) to study the various subdisciplines of science and engineering. There is the need to recognize that some students will enjoy chemistry and others will hate it and *vice versa*, the same being true for the other subdisciplines of science. The same is also true for various sub-disciplines in engineering—love–hate relationships do really exist in the minds of the students.

In fact, many more students could emerge from high school with a real interest in science or engineering if preparation within the school system was better than only adequate. Whether or not a student leans toward or leans away from a career in science or engineering is *student specific*—a fact not recognized by many teachers who often seem to assume that the student will follow their direction irrespective of their own (the student's) preferences. In such cases, the student enters university and takes 1 year or more to "find" himself/herself—that is, it takes the student 1 year or more to declare his/her major subject area for the baccalaureate degree.

Students at the school (prebaccalaureate) level who intend to follow a career in science and engineering need (1) interest, (2) ability, and (3) preuniversity preparation in science and mathematics, but none of these items alone is sufficient. The social standing, the economic standing, the cultural traditions, as well as the gender of any students can influence student interest in science and engineering—not forgetting access to courses, and future educational opportunities. Many preuniversity schools could make greater and more meaningful efforts to encourage students (of both sexes and of *all* ethnic backgrounds) to prepare for science and engineering careers. To do this, the teachers need to understand the interest and abilities of the students and thence to encourage the student along whatever path that s/he (the student) prefers. In fact, many students start thinking about the possibility of a career in science or engineering when their interest is initiated (or sparked) by a high-school teacher or some other adult role model. This is the time when the student should be encouraged and assisted to meet and talk with practicing scientists and engineers, which can be crucial in helping the students to move toward their respective careers.

For students with an interest in science or engineering, it is necessary that they are encouraged to participate in inquiry-based, collaborative learning experiences, which emphasize experimental aspects of science and engineering. In these exercises, the student should learn to (1) observe, (2) collect data, and (3) analyze the data. There is also the need (especially during the preuniversity educational period) for teachers to help students to learn about the relationships among the sciences, mathematics, and engineering subdisciplines. Be that as it may, not all K-12 students receive an acceptable preparation in science and mathematics at the preuniversity level to help them once they enter the hallowed halls of learning (i.e., the university). Many students arrive at the university with what has been *considered* (*alleged*) to be good preparation in science and mathematics (through advanced courses) but which may not have actually helped the student to develop any understanding (let alone a sound understanding) of science or engineering or have helped to enhanced the ability of the student to solve problems.

Every effort should be made to encourage students to undertake a rigorous high-school program of studies, including "advanced placement" courses (courses which offer university-level curricula and examinations to high-school students) where they are available, and which offer an indication of the abilities of the student prior to entering university and also the type of university that may be a fit with the abilities of the student (College Board, 2012). However, there is also the issue of whether or not advanced placement examinations are designed to measure the level of knowledge and understanding about the subject matter of science or engineering (or mathematics) in order to prepare the student for introductory science, mathematics, and engineering courses at the university level.

The need for science and engineering competency on any national basis (Chapter 7) is an opportunity for institutions of higher education to put into place meaningful changes in the ways these subjects are taught. However, if meaningful change is to occur in university admissions policies, the faculty members in the science and engineering departments will have to work more closely with (1) each other, (2) admissions officers, (3) administrators, (4) preuniversity standardized testing agencies, and (5) accrediting organizations to better define specific competencies. In other words, there must be a *specific* plan and not a "willy-nilly mish-mash" of inherent ideas of what should be taught and what is required of the students! Furthermore, high-school science teachers will need to have a deep knowledge of the scientific or mathematical disciplines they teach and detailed knowledge of the strengths and limitations of the scientific method. High-school teachers also will need to be equipped to help students understand that science or engineering is a way of discovery and the acquisition of knowledge (Bransford et al., 2000).

Moreover, teachers also need to keep abreast of recent developments in science and engineering in the interest of staying current for their students as well as for their own benefits. Switching off any mental processes once the teacher has left the school building is not the manner by which current up-to-date in knowledge and teaching methods is accomplished. Many teachers would welcome the opportunity to expand their own level of knowledge, which can then be useful in imparting knowledge to the students. Many professions, though societies and associations (Chapter 3), attempt to provide opportunities for personal enrichment, for increasing professional skills and knowledge, and for interacting with others in the field—as members of a profession (Chapter 1), teachers need and deserve these same opportunities.

Furthermore, preparing scientists and engineers is one of many tasks that teachers at the preuniversity-level schools are asked to do. Instead of encouraging students with the enthusiasm and ability to pursue science or engineering careers, schools (speaking broadly) may often see their function as culling those students (by discouraging them from taking preparatory courses) who do not fit the so-called mold or traditional image of students destined for university. If and when such culling occurs, it is a bias that should cease. All capable high-school students should be made to feel welcome to study science and mathematics (if they choose) and the study of science and mathematics should not be restricted to those students who believe that they *need* such courses for their future careers.

In addition, practices such as *ability grouping* or *ability tracking* may be applied and restrict the preparation of, say, late-developing students who would otherwise be capable of pursuing careers as scientists or engineers. Ability grouping and ability tracking, in which children attend different classes based on their proficiency levels, fell out of favor in the late 1980s and the 1990s as critics charged that they perpetuated inequality by trapping poor and minority students in low-level groups (Oakes, 2005). However, ability grouping has reemerged in many classrooms—a trend that has shown a steady increase since 1996 (National Assessment of Educational Progress,* Washington, DC).

The challenge is to move science and engineering education to completeness in terms of student capabilities (or lack thereof) (NRC, 2003; NSF, 2012b). Schools, therefore, can do a lot to prepare or inhibit students in science and engineering through actions such as course offerings, curricula, tracking, and especially testing. But the teaching of science and engineering leaves much to be desired: (1) The content at the preuniversity level is often woefully inadequate, and (2) techniques for successful lesson delivery in high school need improvement. The latter category often fails because of the use of teaching methods that do not relate the subject matter to the curriculum. For example, a framework for successful lesson delivery in any classroom (preuniversity or university) consists of six parts: (1) gaining the attention of the students and establishing expectations, (2) reviewing relevant, previously learned material, (3) presenting the new information by linking it to previous learning, (4) providing learning guidance or elaboration, (5) providing time for practice and feedback, and (6) providing for spaced practice to enhance retention (ODL, 2013).

Indeed, poor teaching (due to hiring practices and/or the failure to terminate the employment of poor-quality teachers), restricted course offerings, and/or unrealistic science and mathematics curricula (delivered by uncredited teachers) also discourage high-school students from pursuing careers in science and engineering. There are no substitutes for good teachers or for good teaching performance. The education system must continue develop better teachers and offer support in the form of realistic curricula, relevant textbooks, equipment, preparation time, and training—from kindergarten through high school, it is the teacher who inspires or discourages the student. In fact, by the end of the 12th grade (assuming teachers and teaching standards and mentoring have fulfilled the necessary requirements), students should have gained sufficient knowledge of the relevant concepts of science and engineering— sufficient to engage in public discussions on science-related issues and to continue to learn about science throughout their lives. The students should be able to appreciate that the current scientific and engineering understanding of the world is the result of many hundreds of years of creative endeavor and hard work (NRC, 2012).

Finally, mention must also be made here of the concept of *vocational training* (training for a specific vocation in industry, agriculture, or trade), chosen by some students who do not have aspirations to a university education and who may prefer hands-on vocations at the time of leaving high school. The idea is not to categorize every student into a particular research-career path and then educate him/her

---

* http://nces.ed.gov/nationsreportcard/ (accessed August 20, 2013).

accordingly, or to bind students to career paths for which they are not suited. In this respect, the United States would do well to look to some of the systems in Europe that have encouraged and continue to encourage vocational training through apprentice programs.

## 2.3   THE BACCALAUREATE EXPERIENCE

Unfortunately, as has become more common over the past several decades, many beginning baccalaureate-level students or even many undergraduate students are unprepared for a career in science or engineering. This is a serious error of judgment (or lack of judgment) and may be due (in many cases) to a student registering to attend university without even knowing, or even failing to give serious consideration to, the areas of study as well as where the study areas will lead in terms of his/her future. The idea of attending of a student university for a year "to find himself/herself" is an attitude that seemed to evolve during the late 1960s and early 1970s and has continued since then—obviously with some valid or invalid reasons that are not up for discussion or judgment here but certainly are the cause of wonderment and perhaps dismay.

Nevertheless, even before exiting the high-school years and certainly before entering the baccalaureate years, the student should be giving consideration to finalizing his/her career goals and choosing the most appropriate classes to move toward this goal. Attending a broad variety of classes outside of the primary discipline that only *might* be useful at some future time may represent wishful thinking and is surely a waste of time and funds—providing the student has been able to express his/her wishes and has been advised and mentored accordingly. This is where the relevant faculty at the school should have played an extremely important role in the lives and career goals of students. It is also the time for the university faculty to move into the life of the student and play an equally important (if not, an even more important) role in advising students about career goals.

Through the mentor, another effective way for students to learn about undergraduate education is to join (or form) a study group to discuss home assignments and share concerns. The group might decide that, on occasion, the faculty mentor should be invited—not to every meeting but to select meetings where his/her advice may be invaluable. However, the peer pressure from the members of this group on each other, if not directed toward stabling thoughts toward a career, can be as detrimental as beneficial.

### 2.3.1   THE FACULTY

A *faculty* is a division within a university comprising one subject area, or a number of related subject areas. In the United States, such divisions are generally referred to as *colleges* (e.g., in the context of this book, the college of sciences or the college of engineering, or there may be the school of chemistry or the school of engineering). The *faculty* may also be referred to as the *academic staff.* Whatever the title, the cornerstone of any university (for the science-oriented or engineering-oriented student) is the members of the science and engineering faculty. These faculty have been chosen for their individual intellectual excellence but, hopefully in the selection process there was also the requirement (which was not ignored) of the dedication-to-teaching component as well as the commitment of each faculty member to help students develop their potential.

Typically, some faculty members at a university are well known for achievements in their research fields but not always for their ability to teach or their dedication to teaching. There may (should) also be those faculty members whose dedication to teaching outweighs their dedication to research activities. It is not always possible (in fact, it may be impossible) to find such preferences in a printed prospectus, and fulfilling the desire for meaningful information may require a visit to the campus and/or conversations with current and former students. The university student–faculty ratio (specifically in the college of science and the college of engineering as well as in the various departments) should be evident from a prospectus and should be sufficiently low to reflect the commitment of the university to providing students with excellent teacher–student relationships. In an ideal world the name of the professor should be given in the prospectus alongside the courses taught by that professor but, in the real world, this information is not always available until the commencement of classes at the beginning of the semester.

Unless the student is willing to spend an inordinate amount of time investigating the background of each professor—perhaps the high-school teacher can assist in such an endeavor—it is virtually impossible for a would-be university student to select the university based on faculty profiles and research or teaching preferences. The student can, however, investigate the student–faculty ratio, which, assuming that all faculty members are conscientious workers (even when told it is so by the university president and/or by the college dean and/or by the department head, the student can only assume this to be the case), can be of some assistance in the choice of a university.

At the baccalaureate level, the professor must have competence in a wide range of knowledge and skills. The ideal professor must understand the practical aspects of teaching, such as (1) how to set up a syllabus, (2) how to evaluate oral and written work, (3) how to grade fairly, and (4) how to set the pass-fail level and to stay with it even when there are more course/exam fail marks than anticipated. More importantly, the professor must have a broad enough background to be able to teach at all levels of science or engineering and s/he or she must be willing to expand that background by delving into new areas by learning new material and by developing new courses to meet the needs of the students. The professor will (should) know how to select appropriate textbooks (not necessarily books written by him/her, in fact preferably not written by him/her) and teaching materials and how to design and implement creative, interesting teaching techniques for all levels a science or engineering. Of increasing importance is knowledge about the technology that is available to improve teaching, such as video-based language instruction, computer programs, and the use of the Internet—remembering that many students will be following the classroom teaching methods using the now extremely available tablet.

In addition, the professor must have the ability to cooperate effectively with the students as well as with departmental colleagues and peers in other departments. Collaboration in the classroom requires that the professor be able to relate to (empathize with) students and accept them as they are, yet maintain a positive and authoritative (not autocratic) approach to classroom learning. Interested and able students should be encouraged to continue with and excel in the course work but learning will be difficult for the students if the new (or even well-experienced) professor does not understand their needs.

In a university setting, the student will meet with (in addition to the departmental faculty) undergraduate students, graduate students, postdoctoral researchers, and other faculty from which s/he (the student) can gain valuable insights about specific graduate programs, possible careers, and even the current job market. The undergraduate students can become members of the relevant student chapter(s) of a scientific society or an engineering society—such as the American Chemical Society, the American Institute for Chemical Engineers, the Society for Petroleum Engineers, and the Society of Women Engineers, to name only four such societies (Chapter 3).

In addition, there are increasing pressures for university professors to shift the emphasis toward teaching and to assist the students to develop critical thinking skills. Merely educating students to parrot facts without understanding the background or derivation of the facts is to be deplored. Understanding the reasons for the derivation and use of a certain equation is more educational than learning the equation by heart without understanding the whys and wherefores. Unfortunately, this pressure (not to parrot equations or facts) is largely ignored by the university where the faculty members (at least some of the faculty members) are able to attract large research grants. In addition, if the number of positions for permanent faculty has decreased in such a university, senior-citizen faculty members may be able to continue in their position past the retirement age and more part-time faculty members and temporary faculty members (adjunct professors, visit professors) need to be recruited. All of these trends affect the ability of the university to adhere to the original *true mandate* of the university, which is to teach and to assure that university can produce baccalaureate graduates with a full and meaningful education.

Obviously, a challenge facing universities with a strong focus on research is to find a balance between the fundamental activities of teaching and research (Matthews, 2007). Within the scientific and engineering disciplines, attempting to find the flexibility to blend the priorities of teaching and research has been (and continues to be) a perennial (seemingly unsolvable) problem. The standing or reputation of a university is directly related to the research productivity of the faculty, and the competition for grants has led many institutions to place increased emphasis on research at the expense of teaching. Sadly, de-emphasizing teaching in favor of research which, in turn may be de-emphasized in favor of sports programs (Chapter 4) is not the pathway to the collection of knowledge for learning.

While many faculty members may welcome the move of de-emphasizing teaching in favor of research, the essential mandate of the university—teaching—has been lost. In many universities that are or have been heavily oriented to research to the detriment of teaching, research productivity has been given more weight than teaching effectiveness when deciding tenure or promotion. The student may well shake his/her head in wonderment when s/he finds that the quality of teaching is not what s/he was led to expect during the acclimatization (indoctrination or sales) process.

An additional challenge for the predominantly research-oriented university is the need to address the complaints concerning performance of faculty during undergraduate teaching. The faculty member who (1) does not teach, (2) does not attend meetings, and (3) rarely appears on campus is not only an example of unethical behavior toward the students but fails to display any form of loyalty to the student—the paying clients—and to the university, which is the organization paid by the client to provide a service to the clients (teaching the students). Not receiving any assignments from the department head because of his/her actions (or lack of actions) or because s/he (the faculty member) is off-campus doing private consulting and/or travel during a semester, or because of complaints by the students, is surely a reason to build up a paper trail with the goal of changing the attitude of the faculty members and/or terminating his/her employment. It is always a pity (even a disgrace) that the respective board of trustees (or board of regents, or whatever the name of the governing body of the university) does not take up such issues and resolve them as an example to all of the errant faculty who may choose to follow a path of inactivity or personal consulting. Withholding (legally) a pay check or telling the errant professor to find a (paying) job in the outside world can be wonderful stimulus and wake-up call to many inactive faculty members. It is not that some action is better than no action, but a matter of taking the correct action and the exhibition of good governance by the board of trustees (board of regents) (Salmi and Helms, 2013). Furthermore, if the board of trustees makes a wrong decision they (individually and collectively) should be held accountable for that erroneous decision rather than being allowed to withdraw from the decision-making process.

While on the issue of renegade nonperforming professors, during the formative period of the baccalaureate years, science and engineering students should be introduced by one or more of the faculty during lectures to the concept of teamwork—the concept of people working together cooperatively as a team in order to accomplish team goals/objectives (Chapters 3 and 6). This must be viewed as a cooperative effort or a coordinated effort on the part of a group of persons acting together as a team or in the interests of a common cause and unison of the group for a higher cause. The faculty member—often a senior professor sheltering under the umbrella of tenure— who is doing his/her own thing and not adhering to departmental, collegiate, or university goals and policies does not obviously set a good example of teamwork to the student.

Throughout the university baccalaureate experience, the student should have learned and experienced the major roles that the faculty can assume in assisting students to progress through the assigned courses up to the day of graduation. During this time, the student may have evaluated the available job market to make an informed choice (providing s/he is given the real facts by the faculty member), which may also mean dealing with faculty attitudes. Some faculty members and, therefore, their students assign a lower status to nonresearch jobs for people who have earned (or who are about to earn) a doctorate. As a result, postbaccalaureate (doctoral) students who plan for such jobs might be told that they are wasting their education or not living up to the expectations of their respective advisers/mentors (whatever that may mean). That attitude is less prevalent in some professions where nonacademic employment is the norm.

Also, negative attitudes toward nonacademic employment are often less evident during times of job scarcity. Given this scenario, it is necessary for the graduating baccalaureate to remember that there is a wide variety of positions and each position can be as challenging and gratifying as some of the traditional research positions.

### 2.3.2 Teaching Assistants

A *teaching assistant* is an individual who assists a professor with instructional responsibilities. In high school a person who assists the teacher with instructional responsibilities is a *teacher's aide* (TA)—who may be a retired teacher or a young teacher seeking employment. In either case the teacher's aide should have some experience with children in the education system. In the university system, a teaching assistant is a *graduate teaching assistant* (GTA) who is a registered full-time graduate student and has been chosen for the appointment as a result of excellent scholarship and promise as a teacher.

In the context of this book, the teaching assistant should hold a baccalaureate degree in the scientific or engineering field that is closely related to the one in which s/he will be assisting the professor. Under faculty supervision, the duties of a teaching assistant may include: (1) providing help sessions and leading discussion, recitation, laboratory, or quiz sections; (2) holding office conferences with students, (3) preparing materials for faculty-guided classroom or laboratory instruction, (4) assisting the professor to design a course, (5) assisting in the design and/or preparation of exams or quizzes, (6) and proctoring examinations and grading student papers and/or examinations to faculty-guided standards. In actual fact, the professor may pass all of these activities on to the shoulders of the teaching assistant.

If the teaching assistant is selected on the basis of his/her abilities, the basis for the selection *must* be clearly defined as there may be considerable wonderment by the students (the fee-paying clients) about the selection process by which a teaching assistant (or an assistant professor) was hired and whether or not s/he had any prior academic teaching experience—other than laboratory research—and the evaluation of that person during the hiring process. In fact, there seems to be an overall lack of information about the criteria used for hiring university professors at any rank—perhaps even without due diligence in regard to the performance of the candidate—other than research experience with the accompanying success at acquiring research funding.

For a person with limited experience, the teaching assistant might, under such guidelines, be—to coin a well-known phrase or saying—thrown in at the deep end with the hope that s/he can swim. In addition, a considerable number of undergraduate courses in science and engineering are taught by foreign graduate students who do not have a good command of the English language (Matthews, 2007). In fact, rogue or renegade faculty members aside (who do not participate in any departmental activities), many of the complaints from undergraduate students are related to the use of graduate students as teaching assistants in the undergraduate programs, especially in the science and engineering disciplines.

When deciding on a university, the high-school student along with his/her school mentor may recognize the name of an accomplished faculty member and make the decision to attend that university on that basis. There is no knowledge or realization that the accomplishments of the faculty member may be based on research, but the student and the high-school teacher-mentor assume that there is a teaching component that has contributed to the accomplishments, and so the choice is made to attend that university. And then after day one—day one being the first day of classes in the academic year—the student does not see the professor in class but is dismayed to see that the place of the professor is taken by a teaching assistant of unknown capabilities.

Finally, with all of these duties of the teaching assistant taken into account, there may be wonderment about the activities of the professor—that is, the activities of the designated teacher of the course and, presumably, curiosity about the basis on which the course costs are determined. It might also be wondered if this is true value for the fee money paid to the university each year (or each semester) by the student or by his/her parents (the clients).

## 2.4   THE GRADUATE DEGREE EXPERIENCE

A graduate degree is conferred upon a graduate-degree candidate after a period of research in a *graduate school,* which is part of a university—typically, the number of years spent in such research is unspecified at the beginning of the research period. Producing original research is often a significant component of graduate studies, including the writing and defense of a *thesis* (also called a *dissertation*).

A graduate school (a North American term) is that part of a university that awards advanced academic degrees (such as master's degrees and doctoral degrees) with the general requirement that students must have earned a previous undergraduate (baccalaureate) degree. A distinction is typically made between graduate schools (where courses of study do not provide training for a particular profession) and a professional school, which offers specialized advanced degrees in professional fields such as medicine, business, law, and the ministry.

The graduate degree experience has a long been known as a *work-study* program, which takes the form of laboratory research and often (as required by department/ university policies or by the professor/mentor) involves teaching assistantships. This serves the dual purpose of providing training in research—at the same time assisting faculty in their research and teaching and, in the case of teaching assistantships (Chapter 7), teaching responsibilities. Furthermore, just as the baccalaureate student needs a mentor, the graduate student also needs a mentor—perhaps even more so because of the more complex nature of the work and the associated workload.

Thus, upon completion of the baccalaureate degree, the educational system offers the science or engineering student two further degree goals: (1) the master's degree and (2) the doctorate. Both degrees are awarded following a term of *apprenticeship* (study and hands-on laboratory or field work under a supervisor/mentor) in graduate school, but there is a considerable difference between the degrees.

### 2.4.1   THE MASTER'S DEGREE

The baccalaureate degree gives students a basic education in the fundamental knowledge of a chosen scientific discipline (such as chemistry or biology) and engineering (such as chemical engineering or civil engineering). Becoming an independent research worker seeking a higher-level (research) degree demands that the student experience the advanced specialized learning and hands-on apprenticeship of graduate study.

Briefly, the term *apprentice* refers to (1) a person (typically, a young person) bound by legal agreement to work for another for a specific amount of time in return for instruction in a trade, art, or business or (2) a person who is learning a trade or occupation, especially as a member of a labor union, or (3) a beginner or learner. The term is rarely (if at all) used in academia but that is essentially what the scientific or engineering graduate student is: a young person who works for someone else (the professor) in return for instruction in science or engineering.

Apprentice or not, the short course to a postbaccalaureate degree is the master's degree, usually entailing 1 or 2 years of study, mostly in the classroom. The specialized knowledge of the recipient of the master's degree should bring enhanced earning power and professional responsibility. The doctorate requires a longer term commitment, which is a time period of (preferably) 4–5 years (hopefully, not more) of low-paid apprenticeship, which typically gives the survivor full professional standing. However, there are those professors who are quite adamant that to allow a doctoral candidate to leave after 4–5 years of work is unacceptable because s/he (the candidate) is not ready, and that a more prolonged period (7–10 years) is necessary for completion of the work.

A master's degree may entail 2 years of coursework only. Some programs for the master's degree require submittal of a research thesis, while others do not. In the latter case, the master's degree is not so much a terminal degree as a recognition of the coursework (really, an extension of more advanced class work) and qualifying examinations completed after about 2 years in a doctoral program. However, in recent decades, the 2-year master's degree has served in some fields as the terminal degree. For example, the American Society for Engineering Education in 1987 reaffirmed the appropriateness of the master's degree for engineering students not expecting to enter careers in research or university teaching. Almost 5 times as many master's degrees in engineering are awarded each year as doctorates in engineering (for comparison, the same ratio in the physical sciences is close to unity) (NSF, 2012a,b).

The master's degree can serve many purposes: (1) as a professional credential, (2) as an intermediate step to a doctorate, or (3) as a consolation prize for not being able to survive (for a variety of reasons) doctoral study. The master's degree has long been an important final degree for many professions, including engineering and the applied sciences, but the degree is not always suitable for entry into the research workforce. Furthermore, the holder of such a master's degree may find that s/he is (unfortunately) relegated to the ranks of technician within the academic workforce. Thus, the attractiveness of a master's degree in science or engineering varies with the supply and demand of those with higher credentials. In fields with readily available doctoral graduates, the holder of a master's degree may find himself/herself working as an less-than-serious member of a research team or even as a laboratory technician. On the other hand, an active job market (with high salaries) can lure students with a master's degree away from the university and effectively discourage students from continuing to the doctorate level.

## 2.4.2 THE DOCTORATE

Most fields of graduate study in the sciences, as distinguished from engineering, are oriented toward the academic market as well as the industrial job market—somewhat less than half of doctoral scientists work in academic institutions. The doctorate is the basic professional degree in most fields of science, and most science students seek research or teaching positions. Despite growing undergraduate enrollments from the late 1960s to the early 1980s, a stagnant academic job market and slower growth in federal funds for research projects has left many doctoral graduates with the feeling of being underutilized and their qualifications often ignored in terms of task assignments made by supervisors.

A doctorate is appropriate for most students who desire research careers, including in academic research and industrial research. If students are ready to make the leap to graduate school, encourage them to use the telephone, visit campuses (and their home pages), talk with current students and faculty, seek out alumni, attend conferences, and read publications by faculty (NRC, 2003). Personal meetings with professionals and students can bring a feel for the profession and an excellent basis for choosing an appropriate learning environment. In fact, doctoral programs in science and engineering are not only the final formal stage of education, but also represent initiation into the research community.

Doctoral study in the sciences or engineering usually takes 4–5 years (sometimes more)—assuming the student does *not* already have a relevant master's degree. The first year may be partially spent taking advanced classes and preparing for the oral and written qualifying examinations that most universities require new graduate students to pass before they can continue their studies. The beginning graduate student often also teaches undergraduates as a teaching assistant, or may be active in laboratory research. Some entering students have already arranged to work with a certain faculty member, and may have a research agenda planned out. Most new graduate students, however, spend no more than 1 year learning about various research activities at their university.

The choice of a research project and thesis advisor depends on a constellation of factors: (1) positions available in various laboratories, (2) the student's interest, (3) funding opportunities, (4) a mentor's perceptions of what constitutes a significant research problem (the potential thesis topic), and (5) luck (timing and serendipity). Postdoctoral students should be doing laboratory research (nearly) full time after the first year. Moreover, the graduate student is not only a scientist in training, but also a productive researcher. The uncertainty of basic research means that a project *must* (sooner or later) produce meaningful results—sometimes the initial stages of the program is christened *search* rather than *research*—and that research projects are subject to change during the course of thesis research. Above all, to earn a doctorate, the graduate student must (1) make a significant contribution to knowledge in his/her field of research, (2) complete a written thesis, and (3) pass an oral examination.

The typical doctoral program constitutes a two-part system that lasts 4 or more years. The first part consists of up to approximately 2 years of course work, while the second part focuses on a doctoral dissertation based on original laboratory (or field) research that might take 2–3 years or more to complete. The dissertation, which serves as a demonstration of the ability of the candidate to carry out independent research, is the focus of the doctoral program. When completed, the thesis must contain a detailed description of the work performed by the candidate in the form of (1) the actual research work and the results, (2) the relevance of the research to previous work, and (3) the importance of the results in extending and understanding of the area of scholarship. This format is not absolute in all universities and much variation is seen in the content of doctoral theses.

It is customary in most fields of science and engineering for a doctoral candidate to be invited to work as a research assistant (RA) on the project of a faculty member and an aspect of this research project often becomes the subject of the thesis written by the candidate. A traditional expectation of a candidate, as well as of the respective professor, is that the candidate will extend this work by becoming university faculty members. If this is indeed the case, promotion and tenure depend to a great extent on publication arising from the research. But promises made are not always honored.

A properly structured requirement for the demonstrated ability to perform independent research continues to be the most effective means to prepare academically inclined motivated people for research careers. However, original research demands high scientific/engineering and ethical standards, perseverance, and a first-hand understanding of evidence, controls, and problem solving, all of which have value in a wide array of professional careers. In the course of dissertation research, doctoral candidates perform much of the work of faculty research projects and also take on (by choice or as a result of various forms of gentle persuasion) some of the teaching duties at the university. Therefore, institutions and individual professors have incentives to accept and help to educate as many doctoral (and postdoctoral) researchers as they can support on research grants, teaching assistantships, and other sources of funding. By the time a student receives his/her doctorate, the student may have been a research assistant or a teaching assistant. This system is advantageous for institutions, to which it brings motivated students, outside funding, and the prestige of original research programs. In addition, it is advantageous for the doctoral students, supporting an original research experience as part of their education.

Over the last 40 years, the average time it takes graduate students to complete their doctoral programs, called the *time to degree* (TTD), has increased steadily. One measure is the median time that new recipients of doctoral degrees have been registered in graduate school, where many professors consider (in a distinctly uncomplimentary manner) that the student is a source of cheap labor to provide research data for the furtherance of the publication career of the professor. As a result many students now spend 5 or 6 years (with the encouragement of the professor that may often lead to a 10-year period) to obtain the degree rather than a more presentable 3–4 years. At the same time (over the last 40 years), the time required to study for a master's degree does not seem to have increased much beyond from 18 months to 2 years.

The lengthening of the period of graduate work is accompanied by a second trend. It has become more common for new doctoral graduates in many fields to enter a period of postdoctoral study, to work in temporary research positions, and to take a 1-year faculty job before finding a tenure-track or other potentially permanent career-track position. However, *registered time* is the amount of time actually enrolled in graduate school—thus, it might be less than the time elapsed from entry into graduate school and completion of the doctoral work. It is significant that spending relatively more time in doctoral or postdoctoral activities might not be the most effective way to use the talents of young scientists and engineers. Furthermore, because of the potential financial and opportunity costs, it might discourage highly talented people from going into, or staying in, science and engineering.

The median number of years between receipt of a baccalaureate degree and a doctorate in science and engineering has increased from 3 to 5 years during the 1960s to 6–10 years in the past decade. On the other hand, some doctoral students may take a time out between the baccalaureate degree and graduate degree, which can be valuable for gaining work experience (as a scientific intern or an engineering intern) leading to more mature decision-making about careers; thus, an increase in years from bachelor's degree to doctorate can be beneficial.

But registered time to degree (RTTD) has also increased steadily over the last 40 years. The median registered time from start to completion of the work for engineering doctorates increased from 5 years in 1962 to 6 years in 1992. In 1992, it was almost 7 years for doctorates in the life sciences, 6–7 years in the physical sciences, and 7–8 years for the social sciences—and these periods or *terms-of-service* numbers have remained almost constant with only a slight tendency to increase (Kuther, 2013). One finding, reported for psychology, is that the time to degree is longer when there are many students per faculty member or many students overall (Striker, 1994). The Office of Scientific and Engineering Personnel (National Research Council) in 1990 tested a five-variable model over 11 fields of science and could find no causal effects to explain the trend (Tuckman et al., 1990).

Some researchers explain the increase in time to degree by pointing to the increasing complexity and quantity of knowledge required for expertise in a given field. Another possible explanation is the tendency of some faculty to extend the time that the students spend on research projects beyond what is necessary to meet appropriate requirements for a dissertation. In addition, the lack of financial support during the dissertation phase substantially extends time to degree, as do difficulties in topic selection, unrealistic expectations for the amount of work that can be completed in a dissertation, and inadequate guidance by advisers. Still other reasons are (1) poor undergraduate preparation, (2) student reluctance to leave the congenial life of academia, and (3) postponement of graduation in a job market where employment is uncertain. However, there is insufficient investigation (but a lot of speculation) of the reasons for students to spend the extra time that they take to earn a degree—whether in class work, studying for general examinations, doing thesis research, or working as teaching assistants and/or research assistants. In a sparse labor market, students might hope that the extra time can provide them with a better thesis and thus a better chance at a research position.

At the graduate level, the choice of a research adviser is one of the most important decisions a student will make. The mentor can encourage students to shop around carefully, to talk to present and former advisees, and to gain personal impressions through face-to-face interviews. Students should also be advised to examine the performance of possible mentors: (1) publication record, (2) financial-support base, (3) reputation, (4) success of recent graduates, (5) recognition of student accomplishments such as by coauthorship, (6) laboratory organization, and, most importantly, (7) willingness to spend time with students. Much of this information can be learned directly from the potential mentor and from the mentor's current and past students.

Finally, a distinguishing factor of the doctoral degree is the necessity of the degree for employment in academia. Academia is a nontraditional market in its use of tenure and its emphasis on externally funded research, both of which (hopefully) provide stability and insulation from some, though certainly not all, economic incentives that drive the typical labor markets in other areas. Many doctoral graduates who plan an academic career also accept a temporary postdoctoral research appointment following their degree, which provides a valuable period of time for the new doctoral graduate to immerse himself/herself in research, free from teaching responsibilities, so that they can prove themselves as fully fledged independent researchers.

Perhaps more truthfully and to the point, the postdoctorate period is also a labor-market buffer—it is a holding tank (or breathing space) for young researchers during a tight market with few tenure-track academic posts and few jobs in industry but the availability of research money for doctoral programs.

## 2.5   GRADING AND MARKING

In the education system of many countries, *grading* is the process of applying standardized measurements of varying levels of achievement in a course. Grades can be assigned in letters (e.g., A, B, C, D, or F), as a range (e.g., 1–6), as a percentage of a total number correct, as descriptors (excellent, great, satisfactory, needs improvement), or as a number out of a possible total expressed as a percent mark (e.g., a mark of 80 out of a possible 100 is 80%, which is usually referred to as marking).

In some countries, all grades from all current classes are averaged to create a *grade-point average* (GPA) for the marking period. The GPA is calculated by taking the number of grade points a student earned in a given period of time and dividing this by the total number of credits taken. The GPA can be used by potential employers or educational institutions to assess and compare applicants. A *cumulative grade-point average* is a calculation of the average of all of the grades obtained by the student grades for all courses completed during the semester, during the school year, or during a designated period of time.

On the other hand, a marking system traditionally used in many countries involves giving the student a mark typically as a percentage of the possible mark. A critique of this marking system is that most teachers/professors only award a mark between 35% and 75%. This is a completely unfounded criticism, as the same can apply in determining whether or not to award the student an *A* (typically for a mark from 90% to 100%) or a *B* (typically for a mark from 80% to 89%); moreover, marking decisions can be variable between markers. Therefore having fewer marking points, spread more widely, will mean that marking will be fairer, and that a student's best and worst marks will be better represented in their final degree mark. This is a means to retain a system that is not a true representation of the abilities of the students.

Thus, a numerical number is transferred to a letter and then back to a number—this is where the fun and games start and seems to lack logic. Under the grade system, a student gets 90% and this is classed as an A, which is then transferred to a 4.000. But, a mark of 90%, while being an A on the grading system, is equivalent to 3.600 out of a possible 4.000. The student with the 90%-A grade believes that s/he is perfect while there are faults in the work that can be improved on. Then to make matters even more complicated, the numerical equivalents (such as 4.000 and 3.500) of the grades A, B, C, and so on are mathematically combined and averaged to come up with the grade-point average, which might be a number in the order of (hypothetically) 3.776.

So, a mark that is not perfect is given a perfect grade and then reenumerated to get the grade-point average—a number calculated (mathematically manipulated) to three decimal places—that might also signify perfection in the work of the student(s). So the student with a grade-point average of 4.000 may be flaunting his/her superior intelligence (students do flaunt their superior abilities) while they are aware/unaware of flaws in their work—perfection is not omnipresent for each segment of a 4.000 grade-point average.

Then there is the *marking system,* in which a percent mark is assigned for the work by the student(s). At the same time a pass mark is assigned to the course, which means that the student may need a minimum mark to pass the course. Typically the pass mark may be as high as 67% or as low as 40%. In the marking system, the student (if s/he does not already admit to such human flaws) discovers that s/he is not capable of perfection (mother and father may disagree and decide to file suit against the institution) but this is expected, insofar as such legal suits have been allowed and are a way of life for some institutions. The same can happen in the GPA system. In short, the institution caves in to the suit (the attorneys should know better than file such suits but they have to live) and the junior Miss or junior Mr. comes out of this mental and legal skirmish smelling like a rose—perfection has been restored!

Also, in the marking system in institutions of higher education, failure of major components (subjects such as physical chemistry as a part of the baccalaureate in chemistry) can cause the student to fail to receive the degree and s/he may be asked (told) to leave at the end of that particular year and not allowed to register for the subsequent year. But the marking system is also fraught with fun, games, and mathematical manipulation. For example, in many institutions the *performance curve assessment system* (sometimes simply called the *performance bell curve* because of the shape of the curve) is used to determine the pass/fail ratio for a course—or more particularly the number of students designated to pass a course and the number of students designated to fail.

The term *bell curve* (Figure 2.1—in which a plot of the number of students and test grades is shown) is used to describe the mathematical concept called normal distribution, sometimes referred to as Gaussian distribution. The bell-curve shape refers to the shape that is created when a line is plotted using the data points for an item that meets the criteria of *normal distribution.* The center contains the greatest number of a value and therefore would be the highest point on the arc of the line. This point is referred (incorrectly) to as the *mean,* but in simple terms it is the highest number of occurrences of students with a mark in that range.

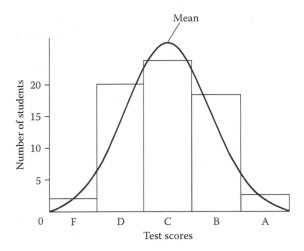

**FIGURE 2.1** Bell curve showing number of students and test grades.

The important aspect of such a curve is that the normal distribution (if such a distribution does indeed exist) of the curve is concentrated in the center and decreases equally on either side. This is significant in that the data has less of a tendency to produce unusually extreme values, called outliers, as compared to other distributions. Also the bell curve signifies that the data is symmetrical and thus we can create reasonable expectations as to the possibility that an outcome will lie within a range to the left or right of the center, once we can measure the amount of deviation contained in the data. These are measured in terms of *standard deviations*.

A bell-curve graph depends on two factors, the mean and the standard deviation. The so-called mean identifies the position of the center and the standard deviation determines the height and width of the bell. For example, a large standard deviation creates a bell that is short and wide, while a small standard deviation creates a tall and narrow curve.

Normal distributions have many convenient properties, so in many cases, especially in science and engineering, variates (student ability and behavior) with unknown distributions are often *assumed* to be normal to allow for probability calculations. Although this is a dangerous assumption, it is assumed (incorrectly) to be a good approximation because any distribution having a finite mean and variance tends to the normal distribution. It is also assumed (incorrectly) that attributes such as test scores follow normal distributions, with few members at the high range and low range and many in the middle range.

While this may make sound sense to some readers, there is another aspect to the bell-curve assessment method. Many teachers/professor do not assign a pass-fail mark and the bell curve is used to calculate the pass-fail mark or pass-fail level for the course. No matter how well or poorly the students have performed in the test or course assignments, according to the bell curve some (perhaps 10%) will *fail* the course and some (perhaps another 10%), because of the shape of the bell curve, will attain excellence. The remainder will vary from *almost excellent* to *average performance* to *almost fail*. However, construction of a real curve based on the true marks for the course will (almost invariably) show a bell curve that is skewed to the low end (Figure 2.2a) or to the high end (Figure 2.2b). Forcing a group of students into a bell-curve distribution of marks and/or grades is statistical disinformation that can lead to grade inflation in which a student receives a grade for course work unwarranted by the level of work or achievement demonstrated. In such a situation, many A and B grades are given and few D and F grades.

On the other hand, if many students are making very low grades on exams, it could be due to very poor instruction, which can be used to signify that teachers/professors care little about their teaching methods. Or there is the teacher/professor who decides that grading and marking methods notwithstanding, s/he is going to ensure that her/his favorites pass and the not-so-favored students fail or all of the class passes—the last option is used for self-gratification to demonstrate the self-proclaimed excellent

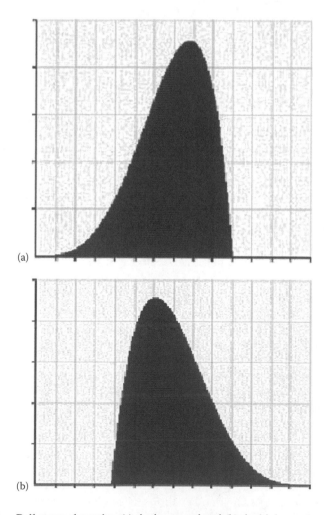

**FIGURE 2.2**    Bell curve skewed to (a) the lower end and (b) the higher end.

teaching methods of the teacher/professor. This is dishonesty in one form, and dishonesty in any form, let alone academic dishonesty, is a serious offense. In the world of academia, dishonesty violates all procedures by giving some students an unfair advantage. But it does not stop with the students (Speight and Foote, 2011).

Using an example cited above (briefly repeated here for convenience and for relevance) let us imagine that a professor (the program professor) heads the MSc-by-course program. The program professor monitors the progress of the students and decides that some of the students who failed the course merit a pass mark, and the professor takes it upon himself/herself to change the marks so that a fail mark for the course becomes a pass. Such actions are untruthful and violate the trust that the professor is given and they render any achievement or recognition based on the cheating to be completely dishonest (Speight and Foote, 2011; Anderson and Kamata, 2013; Bretag, 2013). When this is pointed out to him/her as being unethical, s/he then resorts to a fictitious bell curve to justify his/her changes in the pass-fail ratio.

Teachers and professors need to make themselves more aware of the irregularities that occur when grades or marks are assigned to a student for his/her work. They should also be prepared to report such irregularities—the ethics of grading and marking in science and engineering classes are not only a personal problem but also a collective problem that involves all teachers/professors.

To sum up, whatever system is used there are flaws; some systems have more flaws and are more unrealistic than others. Grading is complicated and subjective, and there may (arguably) be no right or wrong way to grade students. The key is that teachers/professors must be honest in awarding pass or fail marks for a course. It is also arguable that when students receive a good grade it can have a positive effect on their motivation, while poor grades have no motivational value at all. But the overall outcome must be considered when a student who is worth no more than a failing grade is given a passing grade and takes this inability to perform into his/her professional life. The outcome can be disastrous for process chemistry and process engineering when the now former student is given the responsibility of monitoring a high-pressure, high-temperature unit in, say, the petroleum refining industry.

## 2.6  POSTDOCTORAL EDUCATION AND BEYOND

The fundamental purpose of the postdoctoral experience is to extend and deepen the scientific abilities or the engineering abilities of the doctoral graduate, either in the field of the doctorate or a different field. Because postdoctoral positions seldom require full-time administrative or full-time teaching duties, they provide a unique opportunity for researchers to demonstrate originality, creativity, and productivity, which will be primary contributors to their future success in research. In particular, postdoctoral scientists and engineer have the opportunity to produce the lead- or single-author publications by whose quantity and quality they will be judged as they compete for their next professional position.

Scientists and engineers seek postdoctoral experience(s) for different reasons. They may be motivated by the desire to deepen their understanding of a field, to learn a new subfield, to switch fields entirely, or to gain experience in an industrial or government facility. Most postdoctoral scientists and engineer share a desire to enter a career that emphasizes long-term research. Some learn that it is possible to combine research expertise with other skills and find rewarding employment in teaching, consulting, business, law, policy making, and other activities. The postdoctoral years are a time to match educational background, training, and interests with the changing world of employment and to acquire the skills necessary to enter that world.

The decision (usually made during graduate school) about whether to undertake a postdoctoral appointment is seldom easy and should involve consultation with one's adviser and as many mentors or other experienced contacts as possible. A postdoctoral experience may raise the employability of a scientist or engineer as well as being virtually obligatory in certain fields (notably the biological sciences), but a preference (and the associated enjoyment) for research should be the first criterion in choosing a postdoctoral opportunity.

While there are various surveys that purport to show the employment of scientist and engineers, surveys (as well-meaning as they may be) do not always determine the extent to which young scientists and engineers take postdoctoral positions because they cannot find regular employment. One measure of the impact of employment-market problems on the growth of the postdoctoral pool would be an increase in the length of postdoctoral time before a permanent position is found or an increase in the percentage of scientists and engineers who take second or third postdoctoral positions. Another indication would be an increasing percentage of scientists and engineers taking postdoctoral appointments at the institutions where they received their doctorates; this would indicate that professors are retaining their former students as a research assistant when the student cannot find regular a job.

Many such appointments are genuine, above board, not subject to questions, and are a welcome form of employment for many doctoral graduates. However, there are many forms of innuendos and hidden meanings that go with the title research assistant. There may be the graduate student who failed in his/her quest for a doctoral degree and is offered a job by the program professor as a research assistant (read: filing clerk) and then goes on to a successful doctoral graduation where s/he (with the support of the professor) feels there is no harm in bending the previously acquired data to complete a thesis resulting in the award of the doctoral degree (Speight and Foote, 2011). In addition, there may be the well-funded research professor who hires several of his/her students in such positions and s/he (the professor) is then able to show that all of his/her students find postgraduate employment.

Regardless of the proportion of postdoctoral appointees who are in a professional holding pattern, each year they vie with the new class of graduating doctoral students for available positions. The postdoctoral appointees have an advantage in being able to offer more research experience and publications in competing for available research positions. That competition, in turn, increases the trends among new doctorates toward postdoctoral study and nontraditional jobs.

Many scientists with doctorates succeed in moving beyond the laboratory bench to a wide range of careers. Within companies, they might move into marketing, production, manufacturing, sales, or management. Or they can move into such related fields as environmental chemistry, public policy, education, journalism, scientific translation, law, banking, medicine, patent law, public service, and regulation. Biologists at the doctorate level might move to those and other careers, such as biotechnology, pharmaceuticals, biochemical processing, ecology-policy analysis, and patent law.

Engineers, of course, have long moved transparently between academia, industry, and business. All scientists and engineers potentially have the opportunity to use nonresearch skills within science- and engineering-oriented organizations by managing other scientists, developing budgets, and producing plans for new research and development activities (NAS, 1996). Such examples reflect a shift in the conduct of research. Increasingly, the most interesting work is being done at the interfaces between chemistry, biology, physics, engineering, geology, and other subdisciplines of science; this is also the case for the subdisciplines of engineering. That has the effect of blurring the boundaries between traditional disciplines or subdisciplines, so the range of activities in science and engineering is beginning to look more like a continuum than a set of discrete disciplines. The complex challenges of interdisciplinary research demand a broader preparation than does a more traditional disciplinary focus (Tobias et al., 1995).

Postdoctoral study has become the norm in some fields, such as in science and engineering (NAS, 1996, 2000). Some students find that a postdoctoral study in a national or industrial laboratory broadens their outlook and job opportunities and allows them to learn a new research culture. Others find themselves in the position of going from postdoctoral position to postdoctoral position without finding a long-term research position, as well as working for low pay and no benefits for many years. Thus, the decision to undertake postdoctoral work should not be made lightly and should be made only after examination of one's career goals and the career opportunities in that field.

A common problem of postdoctoral students is their lack of institutional connections—they may often feel that they are in limbo in the area known as no man's land. Mentors can help by making the students aware of the nature and location of department offices and by introducing them to other faculty and staff—an obvious step that is often ignored. The mentor can also help by encouraging the department or institution to include postdoctoral students in their seminars, retreats, and meetings with speakers.

Thus, some of the basic obligations that a mentor has to a postdoctoral student are to: (1) help perform research, (2) design a curriculum vitae, (3) rehearse interviews, (4) prepare manuscripts, (5) plan seminars, (6) raise grant money, and (7) learn about the current job market. In addition, a good mentor will maintain sufficiently frequent contact to know about personal or other problems that could hinder progress and will generally make every necessary effort to help the postdoctoral student grow into a mature and productive colleague.

In addition, in return for working on the adviser's project and with relatively low monetary compensation, the postdoctoral candidate has the right to expect good mentoring (Chapter 6) in the form of: (1) oversight, (2) feedback, (3) sympathetic consultation, and (4) periodic evaluation. There should be opportunities to present posters and papers and to learn manuscript writing and grant-proposal writing. The mentor–trainee relationship can be crucial in helping the postdoctoral candidate understand the context of his or her research and the requirements of a career focused on advanced research.

Finally, as a corollary to the postdoctoral experience, it can be noted that female scientists and engineers comprise an increasing proportion of the employed scientific and engineering labor force in industry (Chapter 7) (NRC, 1994; NSF, 2012a,b). While this is due in part to the specific subfields selected by women, a contributing factor to the often-observed decrease in the proportion of women is the attrition rate for female scientists and female engineers in industry, which is double that for men and substantially higher than for other employment sectors. In addition, common recruitment and hiring practices that make extensive use of traditional networks often overlook the available pool of women.

While traditional recruiting and hiring practices were not consciously designed to exclude women, female scientists and female engineers may find that once on the job (in spite of the various affirmative action programs protecting women on the job) they will experience paternalism, sexual harassment, sexual exploitation, allegations of reverse discrimination, different standards for judging the work of men and women, lower salary relative to their male peers, inequitable job assignments, and other aspects of a male-oriented culture that are unfriendly to women (Leach, 2013; Winerip, 2014). Such attitudes add to the discomfort of women in many technical workplaces. In concert with the possibility of an unfriendly attitude toward women in top positions, a major factor that has been cited as determining the size of the industrial scientific and engineering workforce is an individual's selection of a particular degree field, but gender differences still persist (NRC, 1994; Leach, 2013).

Women to a greater extent than men find limited opportunities for advancement, particularly for moving into management positions. The number of women who have aspired to and reached the top levels in corporations is much lower than would be expected, based on the proportion of women in the scientific and engineering work force. This can only be corrected by a conscientious effort to ensure that women (and minorities) receive equal consideration through the implementation of a nonbiased educational system (Chapter 7).

## REFERENCES

Anderson, M.S. and Kamata, T. 2013. Scientific research integrity as a matter of transparency. In: *Global Corruption Report: Education* (Chapter 3.19). Routledge, Taylor & Francis Group, Abingdon, Oxford.

Bransford, J.D., Brown, A.L., and Cocking, R. (eds). 2000. *How People Learn: Brain, Mind, Experience, and School*. National Research Council, National Academy Press, Washington, DC.

Bretag, T. 2013. Short-cut students: From academic misconduct to academic integrity. In: *Global Corruption Report: Education* (Chapter 3.13). Routledge, Taylor & Francis Group, Abingdon, Oxford.

College Board. 2012. *Student Score Distributions AP Exams—May 2012*. The College Board, New York. http://media.collegeboard.com/digitalServices/pdf/research/student_score_distributions_2012.pdf (accessed August 20, 2013).

Curtis, A. 2014. *Report: Wyoming Teacher Training Needs Work*, pp. A1–A8. Laramie Daily Boomerang, Laramie, WY.

Fiegener, M.K. and Proudfoot, S.L. 2013. Baccalaureate Origins of U.S.-trained S&E Doctorate Recipients. Info Brief NSF 13-323. National Center for Science and Engineering Studies (NCSES), National Science Foundation, Washington, DC.

Kuther, T. 2013. What is the Difference between a Master's Degree and a Doctoral Degree? About.com Guide. http://gradschool.about.com/od/admissionsadvice/a/masterphd.htm (accessed July 3, 2013).

Leach, F. 2013. Corruption as abuse of power. In: *Global Corruption Report: Education* (Chapter 2.13). Routledge, Taylor & Francis Group, Abingdon, Oxford.

Matthews, C.M. 2007. Science, engineering, and mathematics education: Status and issues. Report No. 98-871 STM. CRS Report for Congress, Congressional Research Service, Washington, DC.

NAS (National Academy of Sciences). 1996. *Careers in Science and Engineering: A Student Planning Guide to Grad School and Beyond.* The Committee on Science, Engineering, and Public Policy (COSEPUP). National Academy of Sciences, National Academy of Engineering, and the Institute of Medicine, Washington, DC.

NAS (National Academy of Sciences). 2000. *Enhancing the Postdoctoral Experience for Scientists and Engineers: A Guide for Postdoctoral Scholars, Advisers, Institutions, Funding Organizations, and Disciplinary Societies.* National Academy of Sciences, National Academy of Engineering, Institute of Medicine, National Academy Press, Washington, DC.

NRC (National Research Council). 1994. *Women Scientists and Engineers Employed in Industry: Why So Few? Committee on Women in Science and Engineering.* National Research Council, The National Academies Press, Washington, DC.

NRC (National Research Council). 2003. Evaluating and Improving Undergraduate Teaching in Science, Technology, Engineering, and Mathematics. Committee on Recognizing, Evaluating, Rewarding, and Developing Excellence in Teaching of Undergraduate Science, Mathematics, Engineering, and Technology, M.A. Fox and N. Hackerman, (Editors). Center for Education, Division of Behavioral and Social Sciences and Education, The National Academies Press, Washington, DC.

NRC (National Research Council). 2012. *A Framework for K-12 Science Education: Practices, Crosscutting Concepts, and Core Ideas.* Committee on Conceptual Framework for the New K-12 Science Education Standards, National Research Council, The National Academies Press, Washington, DC.

NSF (National Science Foundation). 2012a. Doctorate recipients from U.S. Universities: 2011. Report No. NSF 13-301, National Science Foundation, National Center for Science and Engineering Statistics (NCSES), Arlington, VA, December. http://www.nsf.gov/statistics/doctorates/ (accessed August 7, 2013).

NSF (National Science Foundation). 2012b. Science and engineering indicators 2012. Report No. NSB 12-01. National Center for Science and Engineering Statistics (NCSES), National Science Foundation, Arlington, VA, January.

Oakes, J. 2005. *Keeping Track: How Schools Structure Inequality*, 2nd edn. Yale University Press, New Haven, CT.

ODL (Office of Distance Learning). 2013. *Classroom and Pedagogical Techniques.* Office of Distance Learning, Florida State University, Tallahassee, FL. http://distance.fsu.edu/instructors/classroom-pedagogy (accessed August 20, 2013).

OTA (Office of Technology Assessment). 1988a. Educating scientists and engineers: Grade school to grad school. Report No. OTA-SET-377. Office of Technology Assessment, United States Congress, U.S. Government Printing Office, Washington, DC, June.

OTA (Office of Technology Assessment). 1988b. Elementary and secondary education for science and engineering—A technical memorandum, OTA-TM-SET-41. U.S. Congress, Office of Technology Assessment, Congress of the United States, Washington, DC, December.

OTA (Office of Technology Assessment). 1989. Higher education for science and engineering—A background paper. Report No. OTA-BP-SET-52. Office of Technology Assessment, United States Congress, U.S. Government Printing Office, Washington, DC, March.

Salmi, J. and Helms, R.M. 2013. Governance instruments to combat corruption in high education. In: *Global Corruption Report: Education* (Chapter 3.2). Routledge, Taylor & Francis Group, Abingdon, Oxford.

Speight, J.G. and Foote, R. 2011. *Ethics in Science and Engineering.* Scrivener Publishing, Salem, MA.

Tobias, S., Chubin, D.E., and Aylesworth, K. 1995. *Rethinking Science as a Career: Perceptions and Realities in the Physical Sciences.* Research Corporation for Science Advancement, Tucson, AZ.

Tuckman, H., Coyle, S., and Baem, Y. 1990. *On Time to the Doctorate: A Study of the Increased Time to Complete Doctorates in Science and Engineering.* National Academies Press, Washington, DC.

Winerip, M. 2014. Stepping up to stop sexual assault. *Educational Life, New York Times.* February 9, pp. 14–17.

# 3 Professional Societies and Education

## 3.1 INTRODUCTION

A *professional society* (also called a *professional association*, a *professional organization*, or a *professional body*) is typically a not-for-profit (nonprofit) organization that is registered under the tax code of the United States (or the tax code of the country in which the organization operates), and which has the goal of (1) advancing a particular profession, (2) advancing the interests of individuals engaged in that profession, and (3) advancing or maintaining the public interest in the profession. Briefly as a recap, a *profession* (Chapter 1) is a vocation that is founded upon specialized educational training (Chapter 2) for the purpose of supplying objective (nonbiased) opinions and services to others. Thus, those in recognized scientific and engineering professions provide intellectual and conceptual services in the interest of the client and the public (Jackson, 2010).

An *interprofessional society* is usually a private organization, also registered within the tax code, which has the goal of grouping together participants from all aspects of a scientific and engineering disciplines with the objectives of elaborating policies, guaranteeing equity among the members, facilitating the improvement of the performance of members, and defending the interests of the members. A particular feature of the interprofessional society is that the membership is usually made up of societies and associations that represent the individual professions. Many developing countries have few or no societies that cover scientific or engineering disciplines, and there would appear to be scope for the development of such organizations to promote improved liaison with government, industry, and the public.

By way of further explanation, the definition of a professional society also includes: (1) a group of people in a learned occupation who are entrusted with maintaining or oversight of the legitimate practice of the occupation, (2) an organization that represents the interest of the professional members, (3) a body acting to safeguard the public interest, and (4) an organization that acts to maintain a privileged position as a controlling body. Scientists and engineers may decide that the correct definition is an amalgamation of the first two criteria: (1) "a group of people in a learned occupation who are entrusted with maintaining oversight of the legitimate practice of the occupation" and (2) "an organization that represents the interest of the professional members."

Although the function of scientific and engineering societies is often defined in intimate detail, it may be difficult or troublesome to determine where such societies fit into the education of scientists and engineers. The societies do offer many benefits, including the opportunities to attend symposia and (1) discuss areas of common interest with colleagues who have made important contributions to the field (the *rubbing shoulders syndrome*), (2) discuss any relevant issues with peers on a formal or informal basis, as well as (3) with students who can often be relied upon to challenge assumptions.

With many developments occurring in areas where disciplines overlap, scientists and engineers have many opportunities to meet different people and broaden their expertise. However, it needs to be determined if such symposia are a major (or even minor) part of the education process or if symposia are merely a series of company-paid vacation excursions to the relevant exotic city.

Though professional societies may act to protect the public by maintaining and enforcing standards of training and ethics in their respective professions (Speight and Foote, 2011), they often also act in a manner similar to a labor union (trade union) for the members of the profession—however, such a description is commonly rejected by the administrating body of the professional society. In addition, many professional societies also act as *learned societies* for the areas of scholarship that underlie the scientific or engineering professions.

Unfortunately, there are those societies in which the administrators may tend to forget that they work for the membership and that the membership does not work for the administrators. Therefore, in certain conflicting situations the balance between these two aims may fall in favor of protecting and defending the administrators rather than protecting and defending the professional members. For example, if a society readily publishes (in the monthly magazine or newsletter) a so-called newsy item related to the activities of an administrator but refuses to publish a more pertinent newsy item about a member, the society is failing the membership. In other cases, each monthly issue of the society magazine may be loaded with photographs that show the executive director (or whatever his/her title is) front and center with very little recognition of the members who have received awards. In such cases, the society may be branded as a failure in its goals of representing the membership and the membership should wonder why the society exists.

Initially, the scientist or engineer emerges from the university having (hopefully) been provided with the basics of an education in a chosen discipline or subdiscipline (Chapter 2). But a university may not (some would say often does not) provide the necessary education for scientists and engineers to function in their respective professional areas of scholarship. The young scientist and engineer leaves academia for a world that is largely unknown and seemingly alien, and so, often at a loss for further guidance, s/he will turn to the society for guidance in the professional world, that is, the nonacademic world. The fear of moving into an alien word outside of academia is common among young scientists and engineers and speaks to the lack of mentoring by responsible faculty (or, what is equally to blame, the pretense of mentoring by irresponsible faculty) (Chapters 1, 2, 6 and 7).

Professional societies for scientists and engineers should provide a service that not only involves complete day-to-day administrative management of nonprofit organizations, but also specialized services, including but not limited to (1) a code of ethics, (2) annual and biannual meetings, (3) trade-show meetings, (4) meeting and convention management, (5) salary surveys, and (6) strategic and implementation planning, as well as government relations. The society may also be concerned with activities such as professional development, planned giving, preparing and carrying out a planned-giving program, developing and sustaining membership, and operating educational programs for small and large groups. Not included but perhaps most important of all, the society administrators should listen (or should learn to listen) to the voice(s) of the membership.

Unfortunately, many society administrators focus on the environment (usually a political environment) in which they exist but fortunately the societies are prohibited by law in the United States from engaging in lobbying activities (for the purpose of bringing about changes in the law) at state and federal levels. In some ways, such lobbying activities may even seem to be unethical, but insofar as such activities raise money and increase funding to swell the coffers along with membership fees, the concept that the activities are questionable may be (conveniently) forgotten. A society member may wonder at the number of times a society administrator has claimed to have a friendship with local, state, or federal politicians—but an answer to the question about the friendship and the benefits to the society, if asked, is not always readily forthcoming.

Finally, many societies for scientists and/or engineers attempt to be involved in the development and monitoring of professional educational programs and the updating of skills, and thus perform professional certification—a designation earned by a scientist or engineer to assure qualification to perform a job or task and also to indicate that the scientist or engineering possesses qualifications in the relevant subject area (Duncan, 2013). Sometimes membership of a professional body is assumed to be synonymous with certification, but this is not always the case. Membership of a professional society, as a legal requirement in some scientific and engineering professions, can provide the formal basis for gaining entry into, and setting up a private consulting practice or commercial company within, the profession.

## 3.2   PROFESSIONAL SOCIETIES

The freedom and independence of science and engineering has encouraged the formation of professional societies (learned societies, scholarly societies, and academic societies) to (1) enable discussion of topics of mutual interest among the members and (2) exchange ideas or thoughts that are of mutual interest. In addition, the production of one or more journals by the society as media for the publication of experimental data and opinions derived from the data makes carefully presented knowledge available to the members who read the journals (and whichever libraries subscribe to the journals), thereby disseminating knowledge openly among the membership and also for universal use.

These associations of scientists and engineers are *not* (and were never intended to be) secret societies and it is through meetings of the membership that notice of important discoveries is often made available before the results of the presented work appear in print (Siegelman, 1998). Thus, another valuable function of societies is the presentation of the results of research work to a critical and understanding audience with privilege of discussion by the members leading to education of the newer/younger members. An effect of the presentation and exposure of new work in the light of the knowledge and experience of others is that solutions to unresolved difficulties may be suggested and important implications are discussed that may have been otherwise overlooked. This can lead to a more complete accomplishment of the (project) work to the credit not only of the author but also to that branch of science and engineering.

Another benefit is that an unworthy piece of work may never reach print because of adverse—but constructive and necessary—criticism during the presentation of the work to the society. This is quite common in these days of the Internet, which, unfortunately, does not always contain full copies of prior work (especially work published before 1992, after which the Internet came into more common use) leaving the authors of current work with less than full access to older work. Thus, there is quite an amount of reinventing the wheel because authors do not carry out a meaningful literature search prior to initiating work on a project. For example, the evolution of a research topic initiated in the 1960s with much data presented in journals in the 1970s and 1980s may be missed by the modern researcher because s/he has relied only on the Internet as the source of information. This is, indeed, a major fault of the modern scientist or engineer using only one source (the Internet) as his/her library.

Furthermore, if the presentation of recent work at a society is to be worthwhile, the discussion and criticism from the audience may be frank and severe but it must also be constructive and helpful. It is from such discussions that the newer/younger members are educated in the protocols and decorum of the scientific and/or engineering aspects of society meets. In fact, as a means of educating not only the newer or younger members but also the whole membership, it is worthwhile for scientists to attend sessions related to engineering and for the engineers to attend session related to science. A blending of the two major disciplines cannot hurt but only improve the education of the fertile mind (Chapter 5).

In some meetings of scientific societies and engineering societies, criticism may be of the greatest severity—but may even be complementary—when opponents voice their various opinions and there is no attempt to harm friendship, mutual respect, or future relations. The most futile and disrespectful meetings are those in which cautious members of the audience utter platitudes about "this most interesting and valuable paper," as well as the brilliance of the thought processes of the author, and no one learns anything whatever from the so-called discussion.

In addition, there is the problem of the proponent/opponent who gains attention on the basis of having a question for the speaker but, in reality, proceeds to present a seemingly never-ending monolog on his/her thoughts and ideas that are of no help. The audience members—other than the cronies of the questioner who may have been asked previously (by the questioner) to sit at strategic locations throughout the meeting room to indicate that questions are directed at the speaker from all compass points—generally (but not always openly) condemn such behavior. However, it is part of the duties of the meeting chairman or even the presenter to recognize such behavior for what it is—unethical, self-serving, pompous, and disrespectful—and terminate the questioner, in midsentence if necessary. In such cases, the education of both the neophyte member and the older member has not been enhanced in terms of hearing new data but has been enhanced in terms of learning about unethical behavior.

Then there are those who hesitate to express a frank opinion because they might be judged (by some members of the audience) to be wrong. It is true that each member of an audience may have only a limited experience of the issues under discussion and anything offered in the way of comments may be wrong. But that is the nature of frank and open discussion about the results of scientific research and engineering research, and it is the manner in which other members of the audience are being educated. It is easy to avoid criticism by saying nothing, doing nothing, and being nothing—in such situations, the fly on the wall gets more attention!

### 3.2.1 Scientific Societies

A scientific society is a professional organization for scientists of various disciplines. Some of the scientific societies are umbrella-type organizations which accept many different disciplines, while others are specific to one discipline or specific to one subdiscipline. Many award professional designations, such as *chartered chemist* (in the case of a society for professional chemists). There are also many student-run scientific society chapters, which are associated with the parent professional society, at universities or technical colleges.

Typically, the society is headed by a president who is elected by the membership. Assisting the president in the day-to-day operations of the society there may be the duty of an executive director (or some similarly titled person). The president sits at the head of a committee of other officers (often termed the *board of directors*)—some elected by the membership, others appointed by the president—who are afforded the luxury of an administrative staff (the office personnel) to carry out the general business of the society through communication with the members. Above all, it is the duty of the president and the officers of the society to adhere to the wishes of the membership, unless those wishes are detrimental to the well-being of the society. It is not within the job description of the executive director to decide on policy issues.

It must never be forgotten that the members (and *not* the administrators) *are* the society—an apathetic membership means an ineffective society that offers little education to the membership. Various members may have forgotten these important aspects of society meetings: (1) that an important function of society meetings is to enhance the education of the membership as a whole or to enhance the education of the individual member in the various aspects of science and engineering as well as (2) that these meetings have get-to know-one-another aspects as well, although this does not always happen. As an aside comment, the focus of the society should not be on the number of meetings held or sponsored within a specific time period but on the attendance at the meeting.

Furthermore, *the members are the society* and *the society is the membership*. It would be well if the president and the members of the board of directors as well as the office personnel could remember this statement. Perhaps frequent education programs for the president, the board of directors and the office personnel would be in order, in which administrator–member communication and governance was the focus of the course or workshop (Salmi and Helms, 2013). But it should be recommended that the course (or workshop) be held at the society offices (space permitting, or close-by space not permitting) and not at some exotic resort.

Within a limited sphere, societies exercise their purpose and are a reference for authoritative collective opinion and so, societies administrators claim, the society is of great value. As such, societies need to extend their work beyond the limits set by the administrative body. Proper and adequate education of the membership is often restricted because the society administrators are not interested in such activities and contribute very little to the membership. Because of this, membership education is limited to courses (that *must* be meaningful to the membership) that are associated with an annual or semiannual meeting. Such courses may appear to be popular because of the reputation of the speaker. But the question that must be asked is whether the speaker is a good teacher. More often than not, the answer is a loud, lasting, and resounding "no."

There are at least three functions for scientific societies that are important to track and trace over time and across fields: (1) general education and professional development, (2) prevention and advisement, and (3) enforcement of codes of ethics and handling of complaints regarding members who step outside of the boundaries of ethical behavior. While scientific societies vary in their levels of engagement in these three functions, they differ especially as to whether they are engaged in regulation of scientists within their disciplines.

These three functions are realized through a variety of specific activities, including but not limited to: (1) production of codes of ethics and other formal statements of responsible behavior; (2) professional development workshops; (3) prevention and advising; (4) complaint handling, mediation, and/or enforcement,; (5) development of educational and course materials; (6) providing leadership internal to the field of science, for example, within departments; (7) collaborating across fields of science and education; and (8) providing leadership external to the field, such as national science policy. Scientific societies vary in these categories that are emphasized by the society and in their level of effort assigned to each category. Also, activities can be (1) high profile; (2) low profile; (3) symbolic—that is, lip service only; or (4) concrete—that is, with definitive actions and plans. The degree of attention to the above categories is also an indication that the society administration pays attention to the education of the members. In addition, they can be implemented on a case-by-case basis or be part of a more systemic effort to address academic integrity and academic misconduct (Anderson and Kamata, 2013; Bretag, 2013; Robinson, 2013).

Briefly and by way of definition, misconduct in research is: (1) fabrication of results, which is making up results and recording or reporting them as factual data; (2) falsification, which is manipulation of research materials, equipment, or processes, or changing or omitting data or results such that the result of the research is not accurately represented in the research record; and (3) plagiarism, which is the appropriation of another person's ideas, processes results, or words without giving appropriate credit (Speight and Foote, 2011; Shaw and Despota, 2013).

Beyond observations where specific actions are often absent or where the event may receive mention in the news media, little is known about the role and influence of scientific societies on activities such as guidance, contact with other scientists and engineers, or misconduct in research (Laxminarayan et al., 2000; Speight and Foote, 2011; Anderson and Kamata, 2013). In fact, many society members remain uneducated about this detrimental aspect of science and engineering. For example, in setting forth the parameters of a research agenda, there are several issues that are of paramount importance and which require attention: (1) how to conceptualize research in scientific societies, (2) challenges and complexities in undertaking basic research, (3) strategies for undertaking basic, (4) evaluation of research as the work progresses, (5) assessment of the impact of society guidelines and/or regulations, (6) ensuring that due recognition is given to those scientists and/or engineers participating in the research, and last but by no means least (6) maintaining integrity in research. Since these parameters are not covered in the majority of university courses, these areas are ripe for serious and meaningful society participation (not lip service) in the form of education of the young science or engineer.

Many newly graduated young scientists feel pride in their scholastic achievement (as they should) and are happy with their recently discovered intellectual independence, which has resulted in acquiring new knowledge. However, for those scientists who may be frustrated by a lack of teaching guidance within academia, there are societies that have a strong leaning toward science education. Examples are (1) The Association for Science Education (ASE), (2) The Association of Environmental Engineering and Science Professors (AEESP) (3), The Coalition for Education in the Life Sciences (CELS), (4) The National Earth Science Teachers Association (NESTA), and (5) The National Science Teachers Association (NSTA), all of which offer further education in the various scientific disciplines.

For example, the NSTA, founded in 1944 and headquartered in Arlington, Virginia, is the largest organization in the world committed to promoting excellence and innovation in science teaching and learning for all. The current membership of the association is in the order of 60,000 and includes science teachers, science supervisors, administrators, scientists, business and industry representatives, and others involved in and committed to science education.

As a second example, the National Association for Research in Science Teaching (NARST) is a worldwide organization for improving science teaching and learning through research. Since its inception in 1928, the NARST has promoted research in science education and the communication of knowledge generated by the research. The ultimate goal of the association is to help all learners achieve science literacy.

More scientists with a role in teaching would be well advised to avail themselves of the benefits of membership in such a society (Laxminarayan et al., 2000). In fact, it would not be such a bad idea if these same scientists availed themselves of membership in a society or association that catered to engineering. Gaining extra knowledge (whether it is for teaching or other professional activities) through membership in an engineering society, or membership in an association dedicated to improving teaching methods, would expand the abilities of the scientist (Chapter 6).

### 3.2.2 ENGINEERING SOCIETIES

An engineering society is a professional organization for engineers holding professional qualifications in the various subdisciplines of engineering. Like the scientific societies, some engineering societies are umbrella-type organizations that accept many different disciplines, while others are discipline specific. Many award professional designations, such as *professional engineer* (in the case of a society for professional engineers). There are also many student-run engineering society chapters, which are associated with the parent professional society, at universities or technical colleges.

Engineers, like scientists, have a vital role to play in the development of industrial processes, but the role that the professional engineering societies must play remains undefined. For those engineers who may be frustrated at the lack of teaching guidance, there are societies that have a strong leaning toward engineering education. Examples are (1) The American Society for Engineering Education (ASEE), (2) the National Society of Professional Engineers (NSPE), (3) the Association for Media-based Continuing Education for Engineers (AMCEE), (4) the Association of Environmental Engineering and Science Professors (AEESP), (5) the European Association for Education in Electrical and Information Engineering (EAEEIE), and (6) the European Society for Engineering Education—Société Européenne pour la Formation des Ingénieurs (SEFI), which offer further education in the various engineering disciplines.

For example, ASEE is committed to furthering education in engineering research and in engineering technology. This mission is accomplished by promoting excellence in instruction, research, public service, and practice; exercising worldwide leadership; fostering the technological education of society; and providing quality products and services to members. The ASEE goes back to the basis of the education system by recognizing the role K-12 teachers/educators play in demonstrating exciting engineering concepts to young people. Each summer ASEE convenes a K-12 workshop that gives teachers effective classroom engineering education resources and networking opportunities. In addition, NSPE has endorsed the engineering education legislation (Educating Tomorrow's Engineers Act of 2013) that stresses the importance of taking engineering into K-12 classrooms.

More engineering graduates with a role (or interest) in teaching engineering would be well advised to avail themselves of the benefits of membership in such a society. In fact, it would not be such a bad idea if these same engineers availed themselves of membership in societies or association that catered to engineering education. Gaining extra teaching knowledge through membership in a scientific engineering society or association dedicated to improving teaching methods would expand the abilities of the teacher.

## 3.3   PROFESSIONAL QUALIFICATIONS

A professional is a person who learns and uses his/her skills that were gained during a course of intensive education, typically (in the context of this book but not always) at an institute of higher education (Chapter 1). The general definition of professional qualifications relates to whether or not individuals (especially scientists and engineers) are professionals by examining whether or not they have accepted certain professional values (Parsons, 1954). Furthermore, scientific and engineering professionals generally (1) accept scientific or engineering standards in their work, (2) restrict their work activities to areas in which they are technically competent, (3) use objectivity in their work, (4) put the interests of the clients before their own, and (5) avoid emotional involvement. In addition, scientists and engineers (through education, meetings, and publication of research findings) move their colleagues to accept their findings through persuasion using logic and objectivity and not through plagiarism (Goodman, 1989; Shaw and Despota, 2013).

The latest and a very important area of professional development is the establishment of formal codes of ethics (Chapter 6), which usually include (1) rules to exclude unqualified and unscrupulous practitioners, (2) rules to reduce internal competition, and (3) rules to protect clients and emphasize the ideal service to society. Be that as it may, the new graduate, particularly the new doctoral graduate, will seek some form of further education from the society, some of which may be related to ethical behavior. Too often, such assistance is not forthcoming and the new hopeful member may wonder why s/he only receives contact or communication from the society once a year, when the annual fees are due twice a year if s/he waits for the second notice for fee renewal.

If the young scientist or engineer is looking beyond his/her degree for professional recognition, s/he may have a frustrating search. While the doctorate is recognized in many countries, other forms of professional recognition are not always available. There is, of course, the *professional engineer* (P. Eng.) designation that is based on education and experience and is recognized throughout North America but not always recognized in other countries. For chemists, there is the *chartered chemist* (CChem) designation of the Royal Society of Chemistry (UK)—considered by some to be the nonuniversity equivalent of a master's degree—which recognizes the experienced practicing chemist who has demonstrated an in-depth knowledge of chemistry, significant personal achievements based upon chemistry, professionalism in the workplace, and a commitment to maintaining technical expertise through continuing professional development. This is also recognized in some countries but it is not widely known in many countries.

Also, there is automatic recognition of professional qualifications in the European Union for seven professions that are known as *sectoral professions*: (1) architects, (2) dentists, (3) medical doctors, (4) midwives, (5) nurses, (6) pharmacists, and (7) veterinary surgeons. There are contact points in every country of the European Union that can give a scientist or engineer information on the recognition of professional qualifications (national law, procedures to be followed) and guide the professional through the administrative formalities that need to be completed. The professional can also use the one-stop shopping opportunity provided by the *Services Directive*. However, if the authorities of the host country find significant differences between the training acquired in country of origin of the professional (including postdegree professional experience) and that required for the same work in the host country, the host country may require the professional to complete a training period and/or an aptitude test.

It is time that the various scientific and engineering professional societies in the United States entered into such programs so that any professional designation (other than the university-earned doctorate or other degree) bestowed on the member can be recognized on a worldwide basis.

Finally, the latest area concerning closing a knowledge gap in professional development is the establishment of a formal code of ethics, which usually includes (1) guidelines/rules to exclude unqualified and unscrupulous practitioners, (2) rules to reduce internal competition, and (3) rules to protect clients and emphasize the ideal service to society. A code of ethics usually comes at the end of the profession-alization process but this tends to avoid detailed discussion of "dos and don'ts" and should be an integral part of the learning process (Chapter 4) (Speight and Foote, 2011; Heyneman, 2013).

## 3.4  FURTHER EDUCATION

Postuniversity education (*continuing education, further education*) is generally taken to mean formal courses that are not intended to lead toward a degree but to add to the knowledge of the baccalaureate graduate or the graduate with an advanced degree (Harvey et al., 1993). Even though credit and degrees are not the primary objectives of continuing education courses, one might anticipate that universities would play a major role in such education—universities start the process by being responsible for the undergraduate and graduate education of engineers and they also have the facilities and the faculty for continuing education courses. Therefore, it is reasonable to expect that almost all universities will have significant activities in continuing education. But this is not always the case—a small number of universities (typically in urban centers) with large, well-organized extension programs engage in such programs, but most institutions use their resources for undergraduates, gradu-ate students, and research. Continuing education courses are not always a part of the university psyche.

Supposedly, the function of a university is to serve the public (rich and poor, pro-vided the scholastic aptitude is present), and this includes the scientific and engineer-ing professions, for which universities may (but often do not) provide professional courses. Moreover, if the university does not have further education as a written clause in the charter or does not have an accepted written mandate relating to further education (of the scientific and engineering professions), the university administra-tion and the teaching faculty (or research faculty) may be unwilling to participate in such continuing education activities.

The basic goal of all universities, which should be fully and clearly expressed in the charter of the institution and which should be followed conscientiously, is to educate undergraduate and graduate (advanced degree) students. Furthermore, research activities may not be a part of the university charter or mandate but are usually included in such umbrella statements. In contrast, the specific goals for a continuing education program within the hallowed halls of the university system are usually not as well defined and often come under the heading of public ser-vice rather than education. Be that as it may, the university often does not offer continuing education programs and leaves such activities to the local community college, which may welcome the opportunity to shoulder the burden associated with any form of continuing education and the quality of such programs (Harvey et al., 1993).

Whether planned for or not, the continuing education programs of any university can influence course type and course content, research, and (most importantly) rela-tions with industry. Even though the net flow of subject material may not be from credit courses to continuing education courses, the latter types of courses contribute to the credit courses (NAS, 1985). In addition, short courses provide a showcase for faculty research and often lead to consulting opportunities. Thus, continuing education programs serve as a bridge between industry and academia.

Continuing education students are not compelled by degree requirements to finish a course they find boring or irrelevant or to tolerate a teacher/professor who is not being professional in his/her presentation and demeanor towards the students—many may be at the stage known as midcareer, when many scientists and engineers are seeking (or have engineered, pun intended) a change in career goals. These students challenge the instructor, and also teaching older students can be a stimulating and learning experience for the instructor. However, the professor who boasts (seriously or jokingly) that s/he can present the course (1) with his/her eyes closed, (2) without much work, or (3) that they have not changed her/his course notes for 20 years should (and so likely will) be taken to task by the mature student attendees who are attending the course to learn of the latest developments and expect up-to-date course content.

In fact, as a consequence of the shortage of regular faculty and insufficient incentives or failure to be up-to-date in their respective areas of scholarship, many of the instructors for continuing education courses are recruited from industry or from government. Such instructors are generally good classroom teachers because they usually enjoy this type of work and their reappointment is based on performance—as opposed to the permanent faculty who may never experience a meaningful performance review (Chapter 1) and often do not worry about performance.

The continuing education teachers/professors/instructors that are "on loan" from industry usually have the same academic qualifications as regular faculty but not the record of published research achievements. On the other hand, the instructors from industry and government are up-to-date in their respective areas of scholarship and have a good sense for the applications of scientific and engineering research as well as being well qualified to teach such subjects. Furthermore, contrary to the beliefs and claims of many professors, scholarship *does* continue outside of the university and many industrial libraries are at least the equivalent of a university library in terms of availability of relevant information storage and retrieval.

Thus, one area where professional societies can play a major role is in insisting that educational standards be observed by the membership. A society may formulate a route for further education but if the educational interests of the membership are not monitored, the route to other educational highs is a mere verbal embellishment (in fact, an adornment) and is functionally useless—the phrase "using an umbrella when standing at the base of an exploding volcano in the hope that the umbrella will offer protection" springs to mind. Good luck!

In this light, professional societies need to promote the satisfaction of the joy of further education as well as the exhilaration and satisfaction of discovery that scientific and engineering research offers (Chapter 1). However, attendance by a scientist or engineer at an annual or semiannual society meeting for the sole reason of presenting a status report on his/her project without attending any further presentations is not part of the exhilaration and satisfaction of research. Unbelievable as this may seem, such people do actually exist in academia and also, sadly, in the commercial side of science and engineering.

Postgraduate education also is evident in the form of many important developments that occur in areas where disciplines overlap—scientists and engineers have many opportunities to work with different people, explore new fields, and broaden their expertise. However, education though research can entail frustrations and many disappointments as well as an equal number of satisfactions. An experiment may fail because of poor design, technical complications, or the complete misunderstanding of mother nature by the experimentalists. A favored hypothesis may turn out to be incorrect after consuming many months of effort. Colleagues may disagree over the validity of experimental data, the interpretation of results, or the assignment of credit for work done. Difficulties such as these are virtually impossible to avoid in science and engineering and can seem to be earth-shaking to the novice as well as to the senior scientist or engineer alike. Yet struggling with such difficulties can also offer motivation toward further important progress.

The point of such work is for the scientist or engineer not to bemoan his/her fate or bad luck, or even to criticize colleagues (although they may deserve criticism) but to think about how the work could be done better and successfully. Professors guiding students through an advanced (research) degree often forget that a negative result from an experiment can often mean as much (sometimes more) than a positive result. This attitude that negative results are bad pervades academia and industry and will live forever in the life and career of the post–master's-degree candidate (who really wished to study for a doctorate but had the misfortune to produce several so-called negative results). The professor might be correct when s/he observes that a negative result is not publishable but often deliberately omits to add that there is much that can be learned from a negative result! This is where mentoring often fails and the graduate student remembers the attitude of the professor, which may carry throughout the student's career.

In some cases the experimental methods used to arrive at scientific and engineering knowledge are not very well defined. For example, (1) experimental techniques are often pushed to the limit of the method, (2) the signal is difficult to separate from the noise, (3) unknown sources of error occur and are even plentiful, and (4) even the question to be answered is not well defined. In such an uncertain situation it is extremely difficult (even impossible) to determine which data are reliable when considering a mass of confusing and sometimes contradictory experimental observations. This is where education (or experience—to all intents and purposes, the same thing—and responsible mentoring by the professor) plays a major role.

In any stage of an experimental investigation, researchers in science and engineering have to be extremely clear—to themselves and to others—about the methods being used to gather and analyze experimental data and the ensuing methods of data analysis. Other scientists and/or engineers will be looking at judging not only the validity of the data but also the validity and accuracy of the methods used to derive those data. The development of new methods can be a controversial process, as scientists seek to determine whether a given method can serve as a reliable source of new information. If a scientist or engineer is not forthcoming about the procedures used to derive a new and (hopefully) meaningful result, the validation of that experimental method (and the results) by others will be subject to much skepticism.

Individuals operate according to their own beliefs—including education received over the years—of what should be considered to be the next step and what might be an incorrect next step. There must be some overriding educational research planning moment for scientists and engineers. However there will always be those scientists and engineers whose plan is very simple: "self first, self last, and, if there is anything left, self again." If the experiment does not work, it is the fault of (1) the technician, (2) the indoor temperature, (3) the weather, (4) the color scheme of the laboratory walls, (5) the fly on the laboratory wall, (6) anyone who happens to walk by the laboratory, (7) anyone else who can be blamed, and (8) even the dog and cat at home.

In such a case, the educational process might be considered to have failed such a researcher *or* the researcher might be considered to have failed the educational process. The role of the educational process is characterized by both descriptive and prescriptive aspects. A researcher can choose to affirm or deny role responsibility, particularly when the occupant of a position is a leading and well-known scientist or engineer. Thus, it might be expected that the requisite educational knowledge and skills demanded in these esteemed positions would be sufficient to guarantee research performance, except in a few extraordinary cases (Resnik, 1998; NRC, 2002; Speight and Foote, 2011). This is akin to the futile and disrespectful meetings (mentioned above) in which cautious members of the audience utter platitudes about "this most interesting and valuable paper" as well as the brilliance of the thought processes of the author, and no one learns anything whatever from the so called discussion.

Many professional societies have attempted to develop educational programs that encompass a broad range of disciplines as a means of fostering research performance. These codes presumably represent the ideals and core values of the scientific and engineering professions, and can be used to transmit those values and more detailed educational prescriptions as part of the education of scientists and practitioners. They also provide standards for reviewing claims of "not-me-but-it-was-him" for sanctioning failure in the laboratory.

All programs are supposed to encourage general good research performance, which can be summarized as: (1) conducting and reporting research, (2) giving expert consultation, (3) delivering service, (4) working within the boundaries of educational competence, (5) following all applicable regulations and procedures, and above all (6) being a shining example of the educational discipline. But the sun does not always shine or the shining example may be tarnished!

Thus, the scientific or engineering professional needs (if this has not already been done) to examine the priority of continuing education programs for engineering in light of their role during the coming decade and then make a commitment to meet the needs of the members. In addition, the society should work closely with industry in developing clear objectives for the continuing education of the membership. As a result, the society might be pleasantly (or pleasingly) surprised at the demand for such courses. Also, because of the need to exploit new educational technologies to accommodate the great diversity among the society membership, the society may need to extend the boundaries of available classes, to respond rapidly to changing technology, and to control the costs of continuing education, and they should assume the responsibility (from the universities) for the continuing education of scientists and engineers.

Finally, continuing education courses are scheduled to accommodate the working hours of the attendees. Evening classes are after work and short courses as well as videotaped courses are designed to minimize the time lost from the job. Whatever the schedule, however, it is likely to conflict to some degree with the commitments of any would-be attendee (Cho et al., 2000). Whether or not the attendee is at peak alertness during continuing education classes is an ongoing issue, which can be circumvented by choosing instructors that are motivated to the point of making the course entertaining as well as instructive.

## 3.5   EFFECTIVENESS OF SOCIETY ACTIVITIES

There is the continuing need for professional societies to reevaluate their educational programs and policies. While many societies do this, perhaps they can do more that is relevant to furthering the education of the scientific and engineering membership. Since World War II, the professional societies that have recognized the need and importance for continuing education of the memberships have become a major, highly efficient means of technology transfer.

There should be little (or no) argument with the notion that societies can play a key role in developing initiatives to help further education and promote skilled professional advancement. Yet, society members often acknowledge that a scientific society or engineering society does not ways offer meaningful education programs. Attending annual and semiannual meetings where the focus in the various sessions is often on the members of the " old boys' club" (equality dictates that the words "old girls club" be added here) who occupy the front row and dominate the question period. Such behavior does not offer much in the way of education to the new member except that s/he finds out *who is supposed to be who* within the membership hierarchy of the subject matter of that particular session of the conference.

### 3.5.1   EDUCATION

*Continuing education* is (in the context of this book) an all-encompassing term within a broad spectrum of scientific or engineering postgraduate learning activities and programs. Recognized forms of postgraduate learning activities within the domain include (1) nondegree career training; (2) workforce training; (3) personal enrichment courses, both on-campus and online; (4) self-directed learning such as through Internet interest groups; (5) personal research activities; and (6) experiential learning as applied to problem solving. There should also be an ethics component, which should also be reflected in society activities.

However, societies should also sponsor learning opportunities in research areas for their members, including (1) activities at society meetings, (2) articles in society publications, and (3) the development and dissemination of educational materials, especially examples involving real-life learning experience. If a society is committed to providing quality education resources for educators, parents, students, volunteers, and the public, there should be a means by which such programs are readily accessible. For example, there should be a way that members can share their technical expertise and demonstrate the application of scientific or engineering concepts to support the teaching and learning of science, mathematics, and technology disciplines. One such method is to offer training workshops for members on how to facilitate in-service programs for local teachers to help them introduce hands-on scientific engineering lessons to their students. These lesson plans must be aligned with education standards and need to be easily used in the classroom.

In fact, societies would be well advised to develop partnerships with the appropriate disciplinary departments in colleges and universities to implement these and other educational initiatives.

### 3.5.2 COLLABORATION AND MENTORING

The focus of collaboration and mentoring (as used here, the words are inseparable) is to develop the whole student through personal education, and so the techniques are broad and require the wisdom not only of the mentor but also of the student in order for the process to be used appropriately (Daloz, 1990). The most common collaboration and mentoring techniques are: (1) accompanying, (2) understanding, (3) stimulating, (4) demonstrating, and (5) reaping what has been sown.

In terms of "accompanying," this requires that a professor/teacher make a commitment in a caring way, which involves taking part in the learning process side by side with the student/learner. Mentors are often confronted with the difficulty of preparing the learner before s/he is ready to change—"understanding" is necessary when the mentor knows that what s/he says may not be understood or even acceptable to learners at first but will make sense and have value to the mentee when the situation requires it. There is also the need for "stimulating" the student, as when collaboration/mentoring reaches a high level, learning can escalate—in such cases the mentor may choose to stimulate the learner right into a situation of change in which a different way of thinking or a re-ordering of values is the result.

"Demonstrating" involves making the study concepts understandable and may also involve use by the mentor of a personal example to demonstrate a learning skill or activity—the mentor demonstrates their subject matter by behavior. Finally, in the "reaping what has been sown" category, the mentor must create an awareness of what the student has learned by experience as well as the ability to draw conclusions, which relates to the key issues of what has been learned and the manner in which the learned knowledge is to be used.

The above categories indicate different ways in which mentoring can be applied, but there are two broad types of mentoring relationships: (1) formal and (2) informal. Formal mentoring refers to a structured process supported by the organization and addressed to target populations. On the other hand, informal mentoring leads to informal relationships that develop on their own between the mentor and the mentee. In business, formal mentoring is part of talent management addressed to populations such as key employees, newly hired graduates, high potentials and future leaders. The matching of mentor and student is often accomplished by a mentoring coordinator or by means of an (online) database registry of the abilities of the mentor and the student.

The lesson to be learned is that scientific and engineering societies should work closely together in developing and implementing learning as a way to bridge gaps in the understanding of educational responsibilities across disciplines and professions. Such a bridge can be built through excellent mentoring programs.

### 3.5.3 RESEARCH AND EVALUATION

Individual scientists and engineers do not act in isolation from their professional peers and are encouraged (or they tend) to work in teams (Chapter 6). There is much that a research system (and other forces in the research system) in an academic setting or in a nonacademic setting can do to stress the importance of shaping the professional climate in which scientists and engineers work.

At present, there has been very little meaningful and formal evaluation of the effectiveness of current educational systems and the initiatives that are supposed to improve these educational systems. There is a lot of talk—and an extremely large number of over-verbose poorly written documents as well as well-written documents—that either (1) extol the virtues of the educational system or (2) unashamedly criticize the educational system, with much emphasis on the second category. However, a more rigorous evaluation of education programs is essential if resources are to be efficiently allocated and if scientists and engineers are to have confidence in the educational functions of the schools and universities. Such evaluation must be sensitive to the heterogeneity (gender, ethnic background, and financial background) of the membership of scientific and engineering societies.

### 3.5.4 RETENTION OF SCIENTISTS AND ENGINEERS

Finally, the various professional societies can do much to encourage scientists and engineers and help retain qualified scientists and engineers in their professional appointments. Management can also do much to achieve this goal.

For example, nothing is more discouraging for the fledgling scientist or engineer to sit in an organization meeting (academic, government, or company meeting) and to observe a senior colleague taking credit for his/her idea (i.e., the idea originated by the young scientist or engineer) without mention of the originator of the idea. However, if the idea is not well accepted and even rejected by the meeting attendees, the senior colleague soon reverses his/her story and points to the young scientist/ engineer as the originator of the idea. This is sufficient to make the young scientist or engineer wonder if s/he made the right choice of work organization. At this time, the thoughts of leaving the institute or organization can represent a major mental debate on the part of the young person. Interesting thoughts and mental meanderings of how to get even with the speaker may also spring to mind!

Retention and success data indicate that a higher percentage of women are more likely than men to complete science and engineering degrees, but the retention rates for both genders have room for improvement (Preston, 1994; Cuny and Aspray, 2000; Washburn and Miller, 2005; Hunt, 2010). It should also be noted that retention data from a range of international sources are contradictory and can vary widely between different institutions in the same country (Ohland et al., 2008; Hunt, 2010). Studies of the participation of women in science and engineering indicate that women who enter science and engineering study often have higher commencing tertiary entrance scores than their male counterparts. As a result, the experience of most women undertaking engineering degrees has certainly improved since the 1990s. However, male students may still be more prepared to commence professional engineering careers than female students. Nevertheless, female students are generally more likely to obtain an engineering graduate position than men.

Retention also includes (in the general sense), the presence of female academic staff in scientific and engineering schools as an important means of providing role models and mentors for female undergraduate and postgraduate students (NSF, 2005). Conversely, the lack of female peers and role models can be discouraging for female students (GRG, 2002). The number and percentage of women faculty members in scientific and engineering is very low overall, particularly at the higher levels of full professor.

In the United States, a minority of women make up the tenure-track faculty in engineering and less than one-tenth of the full professor rank (ASEE, 2009). In a more recent report (Kaminski and Geisler, 2012), individual assistant professors (a total of 2966 faculty) hired in science and engineering since 1990 at 14 universities in the United States were tracked from time of hire to time of departure by using publicly available catalogs and bulletins. The results of survival analysis showed that the chance that any given faculty member will be retained over time is less than 50% and the median time to departure is almost 11 years. Of all those who enter as assistant professors, approximately two thirds (64%) were promoted to associate professor at the same institution. Overall, men and women are retained and promoted at the same rate. In mathematics, however, faculty members leave significantly earlier than for other disciplines, and women leave significantly sooner than men, after 4–5 years compared with 7–8 years.

Ongoing issues such as (1) the attitude toward female faculty in the classroom, (2) the structure of academic programs, and (3) poor faculty attitudes often emphasize the prevalent nineteenth-century attitude that science and engineering are highly competitive, masculine domains. While many universities seem to be committed to increasing the number of women pursuing science and engineering, the programs still focus on male-dominated curricula rather than curricula that are attractive to female students. Verbal manipulation (wordsmithery) prevails over structural change in the curricula—ask any academic faculty member what s/he thinks about a certain subject (in this case, the curriculum) and the result could well be a 30-min (or longer) monolog or diatribe (in this case not a learned discussion but an unlearned discussion) in which nothing is said, no point is made, and no decision is reached. The result is that gender divisions in university faculty tend to remain fairly static (Morgan, 2000; Hathaway et al., 2001).

The quest for power also comes to the forefront of a relationship in what is termed *gender harassment*—a form of hostile environmental harassment (Sekreta, 2006). A solution to preventing gender harassment is to focus on systemic means of discouraging such bias. Perhaps this is finally taking root and showing some progress. One very pleasing improvement in student experience relates to the recognition that harassment and discrimination on the basis of gender is illegal in any form (Conefrey, 2001; Leach, 2013). Harassment in the form of intellectual harassment or gender harassment has decreased over the decades because the charges for such behavior, when they move outside of the university, by any university official has moved beyond a lip-service reprimand and a mere slap on the wrist to being classed as criminal activities.

Gender discrimination can also cause low retention of female scientists and engineers (Chapter 7). For example, a recent survey by the Institution of Chemical Engineers (UK) (IChemE, 2013) has raised concern over the loss of skilled women from chemical engineering, with the note that the profession continues to be male dominated. However, new and developing research has identified some of the key factors and indicators that could help predict and prevent those women most at risk of leaving the engineering profession. Similarly, in the United Kingdom, just one in six (16%) of engineering undergraduates are female. In the United States, less than one in five (18%) of engineering graduates are female. The number of women leaving the engineering profession is also concerning, with some estimates suggesting over half of female engineers are opting for alternative careers or other lifestyle choices. In the United States, it is estimated that as few as one in ten (11%) of the people currently working in engineering are women.

Recent investigations focus on some of the issues and have made early attempts to predict behavior patterns that lead to women exiting the profession early. The developing research could influence management approaches, as well employee intervention and retention strategies in the future (Singh et al., 2013). Using a sample of over 2000 women engineers, the research used *Social Cognitive Career Theory* (SCCT) (the reasons why people choose their career) alongside *turnover theory* (the reasons why people choose to leave their jobs or career). The results showed some close relationships between the two theories—SCCT and turnover theory—and may help to explain why so many women fail to enter the profession, or, if they do so, end up leaving it. The research also confirmed the importance of training and development to retaining more female engineers in the workplace. One major point highlighted by the research was the crucial role of training, development, and support. This is not just about making sure that female engineers perform engineering tasks well, but it is also about helping females to manage multiple work-life roles and even managing the political landscape of a male-dominated work environment.

It is clear for the results that human resource strategies within the various workplaces and espoused by the various professional societies need to become more sophisticated (i.e., more realistic with less lip service), that a greater number of female engineers need rightful recognition, and that the engineering glass ceiling must be removed if the profession is to retain female engineers.

Education though mentorship and real encouragement must be the order of the day for the young scientist and engineer. At both ends of the age and experience line, scientists and engineers can educate each other. The older engineer brings experience and organizational know-how that goes with years of service, while the younger colleague brings new ideas and thoughts that can stimulate a valuable relationship.

### 3.5.5 REGULATORY ASPECTS

Originally, any regulation of the professions was self-regulation (and still is to some extent) but with the growing role of government, statutory bodies have increasingly taken on this role, their members being appointed either by the profession or (increasingly, in extreme cases) by government. Proposals for the introduction or enhancement of statutory regulation may be welcomed by a profession as protecting clients and enhancing its quality and reputation, or as restricting access to the profession and hence enabling higher fees to be charged; or else may be resisted as limiting the freedom of the members to innovate or to practice as, in their professional judgment, they consider best. Besides regulating access to a profession, professional bodies may also set examinations of competence and enforce adherence to an ethical code (Speight and Foote, 2011).

In some countries, scientists and engineers may be required by law to be qualified by a local professional body before they are permitted to practice in that profession. However, this is not required in all countries. In such cases, qualification by a professional society is still considered a prerequisite to practice as most employers and clients stipulate that the individual hold such qualifications before hiring their services. Generally, the scientific and engineering professions tend to be autonomous, which means that scientists and engineers have a high degree of control of their own affairs: professionals are autonomous insofar as they can make independent judgments about their work. This usually means that scientists and engineers have the freedom to exercise their professional judgment.

Given the importance of ethics in science and engineering (Speight and Foote, 2011), it should come as no surprise that many different professional societies, government agencies, universities, and companies have adopted specific codes, rules, and policies relating to research ethics (Chapter 4). However, it is essential that the words *autonomy* and *academic freedom* not be translated as a convenient means to circumvent ethical behavior or unprofessional behavior (Arnold, 2002; Robinson, 2013). In addition to government agencies, universities, and companies, professional scientific and engineering societies should also strive to insist that members avoid bias in experimental design, data analysis, data interpretation, peer review, personnel decisions, grant writing, expert testimony, and other aspects of research where objectivity is expected or required. There should also be an avoidance or minimization of bias or self-deception. This should also include protecting confidential communications, such as papers or grants submitted for publication, personnel records, trade or military secrets, and patient records (Chapter 4) (Speight and Foote, 2011).

Professional scientific and engineering societies can also insist (through regulation and suitable punishment for guilty parties) that members remove themselves from any position of conflict (Cho et al., 2000) by disclosing any personal or financial interests that may affect research and decision making.

## 3.6   THE FUTURE

Professional scientific and engineering societies are attempting to keep abreast of technological change and are expanding their efforts in continuing education. However, a society could do much more in designing and presenting professional development courses to their members. A major difficulty in doing so is the lack of solid information on members' needs, the extent of current activities, and similar points.

Many of the necessary courses that are required by nonacademic scientists and engineers are sponsored by the relevant industry or government department. As a means of keeping scientists and engineers up-to-date, the sponsoring department might recruit a retired industrial scientist or engineer (or several such persons) or a university faculty member to conduct or teach the course. The source will be presented to the attendees as a condensed short course in which the material is delivered over a period of five business days. In addition, the attendees have contact with the presenter during breaks and even after hours. Any presenter worth his/her salt will welcome the opportunity to meet with any or all of the attendees during the after-hours time and will also be willing to be available by telephone or (preferably) e-mail for questions and discussions in the post-course period, which can be a specific period—(decided by the teacher/professor/instructor) and be on the order of 3–6 months after the course and is included by the presenter as part of his/her fee for the course.

The benefit of such in-house courses for a company is that the presenter may/should be asked to sign a nondisclosure agreement so that company attendees can discuss current confidential issues with the presenter. This often seems the ideal way for companies to sponsor further education, although they will also send professional and administrative staff to society-sponsored courses. But, whatever the nature of the course, such courses are useful and should continue.

In the context of the present chapter, professional societies can (and some do) fill an important role in meeting the continuing education needs of scientists and engineers. Societies provide the only mechanism available to engineers for remaining up-to-date after completing their formal education. However, several gaps exist in the programs of many societies, and once the gaps have been defined by the membership and not always by the administration, serious efforts should be made to fill these gaps. In fact, one of the issues is that some professional societies may not (or do not) recognize the need to develop alternative education plans for their members and to provide the educational modules and programs necessary to carry out these plans. And there is the need to tailor the course according to the different groups of scientists and engineers.

The simplest way to envision the groups is on the bases of age: (1) mature scientists and engineers—notice the word *older* is not used here, and (2) the younger scientists and engineers. The more mature scientists and engineers are knowledgeable, but they may not readily adaptable to new trends in technology. Also, because these people hold the power positions in the nonacademic organizational structure, changes in programs are not always readily acceptable. On the other hand, the younger scientists and engineers tend to lack the motivation and the means of taking part in continuing education programs. Such continuing education programs, therefore, must be designed both to motivate the more mature group to adapt to changes in methods and to make it possible for the younger members to participate in them.

Furthermore, as scientific and engineering projects become more multidisciplinary, it is essential that a similar multidisciplinary approach be used in developing new continuing education courses. Representatives from industry need to be necessary and effective members of these teams, because industry now recognizes that continuing education is a cost of doing business and not a fringe benefit. One way to strengthen the bridge between industry and a society would be for the society to create the position of scientist/engineer/educator—a person with significant responsibility both as a practicing scientist/engineer and as a teacher/lecturer. Teaching duties could include course development and such an individual within the society could extend contact with industry and academia.

In fact, as a means of helping the membership look to the future, a scientific or engineering professional society can do more to anticipate trends in technology and build them into continuing education programs based on modern delivery techniques. While conventional delivery methods (such as conferences, published conference proceedings, courses, and trade shows) will continue to be necessary and useful, newer methods such as video and audio courses, program tapes, teleconferencing, and the like must be accepted and used to broaden the base of participation in continuing education of members of professional societies.

The future of scientific and engineering societies depends on the ability of these societies to respond to the needs and educational requirements of the membership. Identifying educational learning needs can be challenging—many societies may be aware of such needs but are not cognizant of this fact. The alternative to assessing the learning needs of the membership is a return to the old system whereby educational needs came under learning a "one size fits all" label and the courses were designed and sponsored with no true meaning. However, the membership of these societies has recognized the flaws in the "one size fits all" system and a major aspect of continuous professional education is for the society to be relevant in course presentation and sponsorship.

Assessing the learning needs of the membership can help a scientific or engineering society as well as the individual members. The society can use recommended tutors to help plan curricula as well as balance the needs of individuals and the group needs of all members. Learning-needs assessment is a good way to discover and encourage the membership in postuniversity education.

## REFERENCES

Anderson, M.S. and Kamata, T. 2013. Scientific research integrity as a matter of transparency. In: *Global Corruption Report: Education* (Chapter 3.19). Routledge, Taylor & Francis Group, Abingdon, Oxford.

Arnold, L. 2002. Assessing professional behavior: Yesterday, today, and tomorrow. *Academic Medicine*, 77(6): 502–515.

ASEE (American Society of Engineering Education). 2009. *ASEE Connections*. American Society of Engineering Education, Washington, DC.

Bretag, T. 2013. Short-cut students: From academic misconduct to academic integrity. In: *Global Corruption Report: Education* (Chapter 3.13). Routledge, Taylor & Francis Group, Abingdon, Oxford.

Cho, M.K., Shohara, R., Schissel, A., and Rennie, D. 2000. Policies on faculty conflicts of interest at U.S. universities. *Journal of the American Medical Association*, 284: 2203–2208.

Conefrey, T. 2001. Sexual discrimination and women's retention rates in science and engineering programs. *Feminist Teacher*, 13(3): 170–192.

Cuny, J. and Aspray, W. 2000. Recruitment and retention of women graduate students in computer science and engineering. Report of a Workshop. June 20–21 2000. Organized By: The Computing Research Association's Committee on the Status of Women in

Computing Research. Supported in part by the National Science Foundation Grant No. EIA-9812240 awarded to the Computing Research Association for the study on the supply of information technology workers.

Daloz, L.A. 1990. *Effective Teaching and Mentoring*. Jossey Bass, Wiley, San Francisco, CA.

Duncan, G. 2013. Chemists need professional designation for the public good. *Canadian Chemical News*, 65(4): 13.

Goodman, G. 1989. The profession of clinical engineering. *Journal of Clinical Engineering*, 14: 27.

GRG (Goodman Research Group). 2002. Women's experiences in college engineering (WECE Project). Final Report, Goodman Research Group, Cambridge, MA.

Harvey, L., Green, D., and Burrows, A. 1993. Assessing quality in higher education. *Assessment and Evaluation in Higher Education*, 18(2): 143–148.

Hathaway, R.S., Sharp, S., and Davis, C. 2001. Programmatic efforts affect retention of women in science and engineering. *Journal of Women and Minorities in Science and Engineering*, 7: 107–124.

Heyneman, S. 2013. Higher education institutions: Why they matter and why corruption puts them at risk. In: *Global Corruption Report: Education* (Chapter 3.1). Routledge, Taylor & Francis Group, Abingdon, Oxford.

Hunt, J. 2010. Why do women leave science and engineering? NBER Working Paper 15853. National Bureau of Economic Research, Cambridge, MA.

IChemE (Institution of Chemical Engineers). 2013. http://www.icheme.org/media_centre/news/2013/model-predicts-risk-of-women-engineers-leaving-profession.aspx#. UffAnY3VA1M (accessed July 30, 2013).

Jackson, J.A. 2010. *Professions and Professionalization Volume 3: Sociological Studies*, pp. 23–24. Cambridge University Press, Cambridge, United Kingdom.

Kaminski, D. and Geisler, C. 2012. Survival analysis of faculty retention in science and engineering by gender. *Science*, 335(6070): 864–866.

Laxminarayan, S., Bronzino, J.D., Beneken, J.E.W., Usai, S., and Jones, R.D. 2000. The role of professional societies in biomedical engineering. In: J.D. Bronzino (ed.), *The Biomedical Engineering Handbook: Second Edition*. CRC Press, Taylor & Francis Group, Boca Raton, FL.

Leach, F. 2013. Corruption as abuse of power. In: *Global Corruption Report: Education* (Chapter 2.13). Routledge, Taylor & Francis Group, Abingdon, Oxford.

Morgan, L.A. 2000. Is engineering hostile to women? An analysis of data from the 1993 national survey of college graduates. *American Sociological Review*, 65(2): 316–321.

NAS (National Academy of Sciences). 1985. *Engineering Education and Practice in the United States*. Panel on Continuing Education, Committee on the Education and Utilization of the Engineer, Commission on Engineering and Technical Systems, National Research Council, National Academy of Sciences. National Academy Press, Washington, DC.

NRC (National Research Council). 2002. *Integrity in Scientific Research: Creating an Environment that Promotes Responsible Conduct Committee on Assessing Integrity in Research Environments*. National Research Council, Institute of Medicine. National Academies Press, Washington, DC.

NSF (National Science Foundation). 2005. *The Engineering Workforce: Current State, Issues, and Recommendations*. National Science Foundation, Arlington, VA.

Ohland, M., Sheppard S., Lichtenstein, G., Eris, O., Chachra, D., and Layton, R. 2008. Persistence, engagement, and migration in engineering programs. *Journal of Engineering Education*, 97(3): 259–278.

Parsons, T. 1954. *Essays in Sociological Theories*. Free Press, Simon and Schuster, New York.

Preston, A.E. 1994. Why have all the women gone? A study of exit of women from the science and engineering professions. *American Economic Review*, 84: 1446–1462.

Resnik, D.B. 1998. *The Ethics of Science: An Introduction.* Routledge Publishers, New York.

Robinson, D. 2013. Corrupting research integrity. In: *Global Corruption Report: Education* (Chapter 3.18). Routledge, Taylor & Francis Group, Abingdon, Oxford.

Salmi, J. and Helms, R.M. 2013. Governance instruments to combat corruption in high education. In: *Global Corruption Report: Education* (Chapter 3.2). Routledge, Taylor & Francis Group, Abingdon, Oxford.

Sekreta, E. 2006. Sexual harassment, misconduct, and the atmosphere of the laboratory: The legal and professional challenges faced by women physical science researchers at educational institutions. *Duke Journal of Gender Law and Policy Volume,* 13: 115–137.

Shaw, M.M. and Despota, K. 2013. Journals: At the front line of integrity in academic research. In: *Global Corruption Report: Education* (Chapter 3.20). Routledge, Taylor & Francis Group, Abingdon, Oxford.

Siegelman, S.S. 1998. The genesis of modern science: Contributions of scientific societies and scientific journals. *Radiology,* 2008: 9–16.

Singh, R., Fouad, N.A., Fitzpatrick, M.E., Liu, J.P., Cappaert, K.J., and Figuereido, C. 2013. Stemming the tide: Predicting women engineers' intentions to leave. *Journal of Vocational Behavior,* 83(3): 281–294.

Speight, J.G. and Foote, R. 2011. *Ethics in Science and Engineering.* Scrivener Publishing, Salem, MA.

Washburn, M.H. and Miller, S.G. 2005. Retaining undergraduate women in science, engineering, and technology: A survey of a student organization. *Journal of College Student Retention,* 6(2): 155–168.

# 4 Gaps in Knowledge

## 4.1 INTRODUCTION

A gap in knowledge (or an *information gap*) is a gap where students are, for a variety of reasons, missing the knowledge (information) that they need to complete a task. The knowledge gap may also apply to the ability of students to converse with each other as the conversation pertains to the need to find knowledge or information—whether the knowledge comes from an in-the-hand textbook or from a digital source (Neuman and Celano, 2006; Buitelaar and Cimiano, 2008; Ginsberg, 2010).

Before moving on a discussion of gaps in knowledge, it is necessary to understand the mechanics by which students can fill in such gaps. In this respect, learning from textbooks has attracted considerable attention, resulting in a wide variety of approaches to the extraction of knowledge from textual data. However, the difference between the spoken language on the one hand (many students have never been taught the art of verbally expressing ideas) and formalized textual knowledge on the other is often significant (Buitelaar and Cimiano, 2008). While some students are skilled in the art of meaningful or descriptive conversation, they may find many texts difficult to read insofar as it is not always clear what the author meant when s/he used certain words or phrases. In fact, in the world of academia, the need to continually consult a dictionary while attempting to read a scientific text or an engineering text is not conducive to learning, let alone to filling in a knowledge gap.

When writing a scientific or engineering article or a book chapter or a book, there is a tendency among certain authors to assume that the reader will have a basic background knowledge (some students may be even more advanced in their background knowledge) of the relevant subject areas which are presented in the text, while focusing on a very specific aspect of science and engineering that they (the authors) wish to convey to the readers. Thus, most of the knowledge in such texts is actually implicit and remains under the surface—some would prefer to say that the knowledge is learned by reading between the lines. Furthermore, authors of such works may use language in a rather vague and underspecified way (the *cautious language of the scientist* or *cautious language of the engineer*) which may hesitate to get to the point of the work and never really discuss the conclusions reached by the author in any great detail and, often, without much clarity. The merits of the work and the conclusions must be defined and discussed declaratively and explicitly and be understandable by the reader; otherwise the reader will leave the work with a gap in knowledge insofar as s/he has to wonder what all of the writing meant. This requires a fundamental knowledge of the language of the text—the poor writing of the text notwithstanding—and the ability of the reader to decide if the text is useful or not at all, and if the efforts of reading the material are a worthwhile expenditure of time.

Early in the career of a scientist and/or engineer, the importance of identifying the usefulness of a text and whether or not the text will assist the reader to fill in any knowledge gaps is an important factor that must be addressed during teaching. This requires careful and assiduous teaching in the various aspects (or vagaries) of language and reading, which should have been introduced in the school system—typically it is too late to introduce such techniques in the university system, even though there may be university students (at an age of more than 18 years) with a reading age of a 6–8 year old student in the school system. If the various language and reading skills have not been taught (instilled into the student) as part of their preliminary education, there will be knowledge gaps no matter how well or how assiduously the student studies. Instilling the various language and reading skills should be an important part of preparing the student to fill in knowledge gaps for the remainder of his/her preliminary education, secondary education, university education, and postuniversity education. With this basic teaching, the student should be equipped to recognize knowledge gaps and take the necessary steps to fill these gaps. Without these basic skills the future of the student is suspect.

Thus, as it pertains to science and engineering as well as other disciplines, the education systems of schools may appear (with some justification) to be more complex and demanding while the administrative organization of schools appears, for the most part, to have remained static (if not rigid)—"we have always done it this way" is a prevalent attitude among some teachers and school administrators. Such an attitude may be one of the hurdles in the early education of scientists and engineers. In fact, if students are expected to attain high standards of academic performance, teachers in schools should be held accountable for ensuring that students moving toward a career in science or engineering—in fact in all disciplines—should be able to meet high educational standards. Furthermore, the prevailing assumption is that teachers should know most of what they need to know about how to teach before they enter the classroom (Elmore, 2002). Moreover, it is essential that a teacher has accreditation in the subject that s/he is teaching through a professional organization (Chapter 1) (TEAC, 2014). If not, there must be serious action taken to either ensure that the teacher is given the chance to get accreditation or s/he should be removed from that classroom. The same principles can be applied to a university professor where the professor might only be one lecture ahead of the students attending the course. In other words, knowledge gaps often commence because of the inability of the teacher/professor to pass on knowledge to the student. And an accredited teacher/professor who teaches by standing with his/her back to the class (the students) writing meaningless equations (or other forms of meaningless expressions) on the board is an equal failure in terms of imparting knowledge to the students.

In filling these knowledge gaps, schools and universities *must* have in place accredited, knowledgeable, competent, and honest teachers as well as a series of organizational processes that help the transmittal of knowledge to the students. In fact, knowledge should not be viewed as an object or as a resource to be packaged and transferred without follow up but should be shared among teachers/professors and students. Both the deployment of already existing knowledge and the creation of new knowledge are based on processes of interactions between the teacher/professor and the students, which allow information sources to be pulled together to fill the knowledge gaps.

Gaps in knowledge exist when too little is invested in primary-school education, which acts as the base for the entire education system (Stiglitz, 1999). However, higher education may be equally important, not only for closing knowledge gaps but also for playing an important role in bridging knowledge gaps and must benefit more than a small elite group of the population. The stagnating development of all levels of education affects the creativity of students and people in that society.

In fact, education in schools and universities is leaving a period in which questions of practice and its improvement were essentially pushed into the classroom, where doors were shut and teachers/professors were left to develop their own individual ideas and practices, largely ignored by the organizations in which they worked. The next stage of development in any education system, provided there is performance-based accountability, requires the development of a practice of continuous educational improvement—a body of knowledge about how to increase the quality of teacher/professor practice and improve student learning in classrooms, schools, universities, and, in fact, throughout the entire education system (Glencorse, 2013; Stensaker, 2013).

At the core of the education system is the need for professional development, that is, the process of teacher/professor education for the purpose of improving student achievement. The practice of professional development should embody a clear model of teacher/professor learning (Elmore, 2002). The good news is that school teachers often participate in professional development and learning through summer schools and workshops. The bad news is that a professor might travel to many countries to attend technical scientific or engineering symposia to present summaries of his/her research work, but very few ever attend professional-development courses or workshops to improve or even learn teaching skills.

In the area of professional development, it is necessary to define explicitly the new knowledge and skills teachers/professors will learn as a consequence of their participation, how this will be manifested in their professional practice, and what specific activities will lead to this learning. In addition, as part of attending a professional-development course, the attendees (teachers/professors) should understand that the course is not a means for the teacher/professor to: (1) add a line to a résumé, (2) have an enjoyable weekend, if it is a weekend course, and/or (3) renew friendships with dining/drinking cronies. The reason for the course is dedicated to the very serious business of ensuring that the teacher/professor is qualified to teach and has the wherewithal to impart knowledge to students.

In addition, professional-development courses should be designed to stimulate teachers/professors (1) to work collectively on problems of practice within their own organizations as well as with teachers/professors in other organizations, and also (2) to support the knowledge and skill development of individual teachers/professors. Thus, the mission and goals that shape professional development should reflect a path of continuous improvement in the various aspects of student learning *and* retention of knowledge. This does not mean that the student should merely remember facts so that s/he can chose the correct answer from six possible alternate answers, but has the knowledge to write a meaningful essay on the pros and cons of a scientific or engineering issue. Hence, successful professional=development courses should involve work with individual teachers/professors or small groups around the observation of actual teaching of students.

Finally, a successful professional-development course that is designed to improve student learning and the retention of knowledge should be evaluated continuously and primarily on the basis of the effect it has on student achievement by preventing gaps in knowledge (Chapter 7).

Without suitable and continued professional development of teachers and professors, a knowledge gap can be treacherous as it pertains to the future of the student and can be equated to fumbling in the dark. This poses a unique problem to teachers and professors who are not certain if the students in the classroom are gaining any knowledge, which results in the teacher/professor struggling to help the student gain an understanding of the classroom subject matter. Moreover, if there is a belief that the students retain knowledge merely based on the completion of a course prerequisite, the teacher/professor has committed a faux pas and is basing his/her teaching on erroneous assumptions of knowledge retention by the students.

To move beyond such false assumptions and instead of going into the classroom with an inflexible plan (such as "this is what we must cover today no matter what happens") the plan must be flexible so as to cater to the needs of the students. The teacher/professor must be student centered (rather than self-centered) so that s/he can adapt to the needs of the student(s). Furthermore, the teacher/professor must have a very clear goal in terms of the knowledge s/he expects to impart to the students. In order to adapt in such a manner as to help the students, the teacher/professor should open the classroom work with clear statements concerning what the subject matter of the class will be for that period or day. A simple enquiry by the teacher/professor to ascertain/determine what the students already know about that topic or (more pertinently) what the students think that they know, which will vary over a wide range, will allow the teacher/professor to determine the best way to proceed. And this is not based on the erroneous assumption by the teacher/professor but on specific feedback from *all* of the students in the class. In this manner, the teacher/professor can allow the students to lead him/her to discover whether the information from the students is right or wrong; then s/he can determine the obvious gaps in knowledge and the means by which these gaps can be closed for the benefit of the students.

The importance of identifying and filling knowledge gaps has been recognized to be an important factor in student education and cannot be overemphasized. Educational institutions must have (or put) in place a series of knowledge-delivering processes which lead to acceptable levels of education in students. The deployment of already existing knowledge and the creation of new knowledge are based on processes of interaction that derive from the interplay between the students and the teacher/professor. In particular, the focus should be on the processes of knowledge delivery at the various levels of education, through which students will be able to gain new knowledge and explain their ideas and to assemble information to fill knowledge gaps.

## 4.2  PREUNIVERSITY EDUCATION

As students are preparing for the next step of their education after high school, decisions need to be made, and these decisions will play an important role in determining the university to be selected as a follow up to their secondary education experience.

Often enrolling in a preuniversity course may be admirable, if not essential, to assist young adults to prepare for the future.

A *preuniversity course* (PUC) is an intermediate course that is typically of 2 years duration and is usually conducted by state authorities in many countries—the institution of preuniversity education might be called a *community college* or a *technical college*. Any person who is short of the necessary university admittance qualification and desires admission to any university has to pass this course. The course can be considered a bridge course to prepare students for university education—in reality the course is necessary for the student to close any knowledge gaps before entering to university and reading for a baccalaureate.

As a student (and his/her parents) go through the process of making this important decision in relation to an academic future, every avenue must be explored and a preuniversity course may be an option that cannot be ignored. Attending such a course could be the difference between receiving the coveted acceptance letter for a first-choice university or the less-welcome letter that wishes the student good luck as s/he pursues entry into other universities.

There are a number of ways that high-school (or preuniversity) students can explore their career interests and begin preparing for a career in science or engineering. Selecting certain types of middle- and high-school course work and participating in programs and projects can guide and prepare students for university and the rest of their lives. For example, the most successful scientists and engineers must know more than how to create new technologies—they also know how to effectively communicate with others. These professionals generally work closely with teams of other scientists and engineers, management, and/or clientele throughout the development of a product or technology (Chapter 6). Therefore, it is essential that high-school students work on building strong interpersonal, writing, and communications skills before attempting to enter university and, eventually, the workplace.

In fact, administrators and teachers in a preuniversity institution should work toward relevant science and engineering curricula in which subjects taught may (should) not only include: science, integrated science, and computer studies but also language, literature, and ethics in education in science and engineering, which can all serve as a powerful stimulus to raise the quality of school education for prospective scientists and engineers who attempt to enter a university.

A significant part of this endeavor is the role of libraries—each library must house a collection of carefully selected books in order to meet the requirements of the science and engineering (mathematics) courses. The library should contain books related to mathematics, physics, chemistry, and information technology. In addition, it is often preferable to include current newspaper items that address issues in science and engineering as part of the classroom teaching, so as to activate the inquisitive minds of the students.

While all such endeavors made to enter university represent a commendable effort on the part of the student, one must also wonder about the presence of the knowledge gap that has evolved during the education of the student. Schools and other preuniversity teaching institutions might offer that the student was not up to the education level required for a variety of reasons. True as this might be in some (but not all) cases, the institution must also share (a large or small) part of the blame for not recognizing the needs of the students and the requirements to enter university.

## 4.3 UNDERGRADUATE EDUCATION

The transition from high school to university (if necessary, via a preuniversity teaching institution) will bring to the student the immediate realization that s/he has a sizeable knowledge gap, not just from the perspective of areas of scholarship and knowledge but also from the perspective of behavior. Also, there is so much to learn about the new surroundings that the students may not be concerned about aspects relating to the major subject and resulting career. Higher education can last a long time and lead to a long career, so the student will need to decide on a career path. However, it is advisable that the decision on a career has been made before entry into the hallowed halls of learning and that the first year or (or two) in the university system is not the time for the student to find himself/herself.

In many cases, one of the first steps in the application process to enter a university or institute of higher education may be the entrance examinations. To increase their chances of success, some students complete various steps in preparation for exams such as the SAT (variously called the *Scholastic Aptitude Test* or the *Scholastic Assessment Test* as well as other names) or ACT (originally an abbreviation of American College Testing), and passing of these examinations indicates (or does not indicate) the increased likelihood of matriculation (Plank and Jordan, 2001; Balf, 2014).

The extent to which students prepare for entrance exams may be associated with the number of applications made by the student (while in high school) to various postsecondary educational institutions. This decision may also be related to the information available to students at school, at home, or in the community. Students who consult various sources—such as parents, teachers, guidance counselors, and the individual educational institutions of interest to the student—for information about a postsecondary institution have a greater tendency to attend a 2-year or 4-year educational program than those who do not seek out information from these sources (Horn and Chen, 1998; Plank and Jordan, 2001; Hill, 2008; Bettinger et al., 2009; Engberg and Wolniak, 2010). Furthermore, there is a benefit to participating in college awareness programs that offer students such assistance as providing academic support, career development, financial-aid resources, and opportunities to visit campuses.

Another important aspect of decisions by students about and predispositions toward postsecondary enrollment is their perception of costs and access to financial aid (Grodsky and Jones, 2007; Horn et al., 2003; Pérez-Peña, 2014). Informing students about the availability of financial aid and offering assistance with the process of applying for aid have been associated with increased postsecondary enrollment rates and financial-aid applications (Bettinger et al., 2009). Financial aid and low expenses have been cited as important factors in the postsecondary education selection process (Ingels and Dalton, 2008; Engberg and Wolniak, 2010).

Once the entrance exam has been completed successfully, entry into a university system is a big step after leaving high school. First and foremost is the choice of a major area of scholarship—the university may make this a criterion for taking the entrance examinations. Entering a university with the object of finding oneself (whatever that means) is the wrong choice and students entering university would be well advised to have already made the choice of a career path. Choosing a faculty and major prior to university entry will remove some of the uncertainty of leaving home—parents will not be there to look over a student's shoulder to make sure that s/he has done her/his homework. The student now has total responsibility for his/her education.

For most new students, entry into a university community is a culture shock; the student is no longer in the daily company of mother and father. Diversity is a major part of campus life, with people representing all backgrounds, religions, ideologies, and ages. And there are other duties that require the attention of the student. For example, it is the responsibility of the student to (1) ensure that tuition fees are paid on time, (2) register for classes, (3) attend lectures and laboratory classes, and (4) hand in assignments by their due date—no will chase after a student for an errant written assignment that was due a week ago.

In addition, students enter the university education system with a range of prior knowledge, skills, beliefs, and concepts that significantly influence what they notice about the environment and how they organize and interpret it. This, in turn, affects the abilities of the student to remember, reason, solve problems, and acquire new knowledge (Bransford et al., 2000). All students and (professors) should remember that new knowledge is built on existing knowledge. Recognition that new knowledge cannot be built effectively on a weak foundation is essential—building a house on a foundation of sand or a house of cards are analogies that spring to mind. Thus it is important to determine where students' prior knowledge is *fragile*, that is, where it contains inaccuracies, naive assumptions, and/or misunderstandings of the contexts and conditions in which to apply particular skills.

For the would-be scientist or would-be engineer, there are a number of ways to assess knowledge. The student can determine if the knowledge s/he gained in high school is an immediate precursor to the new knowledge presented at university. On the other hand, the professor—having the same question in mind about the level of knowledge of the new students—may administer a simple diagnostic pre-test during the first week of class, which (if the text is well designed) will identify areas of robust or weak understanding. Even if the system of education in the schools and universities falls under the umbrella of a national system, the quality of teaching in the high schools may have led to gaps in the knowledge of the new students. After the results of the pre-test data have been analyzed, the professor can adjust the course content or teaching habits accordingly. In addition, if a small number of students lack the necessary skills for the course, the information from the pre-test can help the professor advise them appropriately, perhaps to seek outside tutoring or even, in extreme cases, to exclude the class from the student's curriculum (i.e., to drop the class).

With respect to science and engineering, the subject matter taught at high school may (or is often likely) not of the necessary caliber for the new student entering university. The knowledge gap may be too wide for the student(s) to adjust. If this is proven to be the case, the high school should be informed immediately of this issue and/or the university should consider a course that will help bridge the knowledge gap between high-school and university courses. Such an occurrence indicates major flaws in the system and that the students need serious help.

However, by the time the science or engineering student reaches maturity and graduates with a baccalaureate degree, any such gaps in knowledge should have been removed and the student should be on a par with other students on a national basis.

## 4.4   POSTGRADUATE EDUCATION

One gap in knowledge in learning science and engineering can occur when the student decides on a research career and realizes that a graduate degree is necessary. At this time, the student may have to choose between (1) remaining at his/her *alma mater*, (2) moving to a completely different university, and (3) moving to an industrial research laboratory or to a government research laboratory or to a laboratory where further education is encouraged.

Whatever the choice, by this time the student should have realized that experts can disagree and facts can contradict one another. To students on this step of the educational ladder everything becomes a matter of perspective and opinion, with all opinions being accorded (or should be being accorded) equal validity. The student should feel more empowered to think for himself/herself and question the perceived wisdom, but may not be able to evaluate the different perspectives or marshal evidence to support his/her own perspective or interpretation. The student may also view the evaluation of his/her work by the professor as being purely subjective. This is not a knowledge gap but is, however, part of the learning process.

During the early 1950s, policies for the support of science and engineering were influenced by the Cold War with a decrease in interest when the Cold War ended—if indeed it ever ended! Subsequently, science and engineering education has gone through various reorganizations and new methods of knowledge production have been initiated—postacademic science or postacademic engineering. As a result, two main assumptions concerning the progress of the university of the twentieth century (which has spilled over into the twenty-first century) are no longer revered: (1) that teachers in the sphere of good undergraduate training have to be good researchers and (2) that all citizens pursuing a university education should receive the same training. The second assumption has been challenged on the basis that a minority (<10%) of the global student population becomes researchers and the policy formulators are not convinced that the economic logic for such a small percentage of the student population is worthwhile. In fact, in the past several decades science and engineering has witnessed a reassessment of the system of the production and reproduction of knowledge, raising the distinct possibility (or realization) that the knowledge involved in all baccalaureate science degrees and that in all engineering degrees are not the same—not only throughout the country but also throughout the different regions of the country.

In the current world, gaps in knowledge are causing considerable emotional concern over a particular topic—*global climate change.* This used to be called *global warming* but the name has been changed because of aberrations in the so-called warming trends of the Earth and the causes thereof. This issue has also been fouled by issues of ethical behavior on the part of some investigators, which is not the subject of this book but has been presented elsewhere (Speight and Foote, 2011).

Relative to the point of the preceding paragraph is the point that such questionable actions created a knowledge gap in the minds of many students as well as in the minds of other researchers. In addition, as is the case with many scientific and engineering projects, both sides of this issue have not been debated fully. Emotions run high—fueled by some misleading articles in the news media—and the knowledge gaps are never really filled with facts. For example, the point that seems to be forgotten (or ignored) in the climate-related debates and publications is that the Earth is resilient to changes and also is currently in an interglacial period. As a result— surprise, surprise!—the temperature of the Earth will increase, making additional knowledge essential to the core of the argument/discussion (Goreham, 2013). The actual extent of the temperature rise is unknown—who was around to measure the temperature increase during the last interglacial period?—but the effects of the interglacial period will make a definite contribution contribute to the overall temperature rise. Perhaps the scientists who ignore such a phenomenon are also guilty of failure to teach their students (and the public) the relevant facts.

This is a specific example where a gap in knowledge (or failure to acknowledge other information that is equivalent to a gap in knowledge) has an effect on the outcome of the root cause of a scientific/engineering issue, in this case global climate-change mitigation. Acknowledgment of additional data, modeling, and analysis could narrow the knowledge gap, and the resulting improved knowledge and empirical experience could assist decision making on climate-change mitigation measures and policies.

In addition to the failure to admit that the Earth is in an interglacial period, there are still gaps in accurate and reliable emission data by sector and by specific processes, especially with regard to the emission of greenhouse gases from various sources, such as deforestation, decay of biomass, and peat fires. Because of the knowledge gaps, consistent treatment of greenhouse gases other than carbon dioxide in the methodologies underlying scenarios for future greenhouse-gas emissions is often lacking in logic and, sometimes, authenticity (Goreham, 2013).

This leads to a deficit in the amount of empirical evidence on the magnitude and direction of the interdependence and interaction of sustainable development and climate change, of mitigation and adaptation relationships in relation to development aspects, and the equity implications of both—there is truly a knowledge gap! Another important gap in knowledge is the information on spill-over effects—the effects of domestic or sectoral mitigation measures on other countries or sectors.

The problem as it relates to the current system of education is the means by which students (especially higher-level baccalaureate or doctoral students) can proceed to plug any such knowledge gaps (Weimer, 2009). In fact, without due care and attention from the teachers/professors, the difference in knowledge levels between education groups is predicted to increase over time, irrespective of the evolution of digital media (Yang and Grabe, 2011). If a student feels that a teacher/professor has a bias, s/he (the student) must make a habit of constantly learning new things. Failure to do so means that the level of knowledge of the student will become dated and will merely follow lines of investigation that cannot be substantiated by current and/or alternate facts, and students may seek alternate means of proving a theory (Speight and Foote, 2011; Goreham, 2013). In the context of this book, it appears that the resolution to the issue of the failure of teachers/professors/researchers to deliver the full and knowledgeable message of any scientific or engineering issue to students is to end the large research university and replace it with quality teachers. At the same time, effective research strategies should be imparted at a research-oriented university.

At all levels of the education system (high school and university), it is the responsibility of the teacher/professor to help the students to grow intellectually by filling on any gaps in knowledge. It is important that the students be provided with challenges that move them gradually out of a false comfort zone so that they can grow and not become discouraged. Learning to ensure that gaps in knowledge do not exist can seem to be a threatening task, and facing too much of a challenge can cause students to rebel or retreat—too little of a challenge, however, and the students will not progress and knowledge gaps remain.

Finally, the education of scientists and engineers in many ways is only beginning when the students receive their degrees and join the work force (NRC, 1985). The direction of a scientific or engineering career may change from time to time, but even if it changes very little, the technology with which it deals is changing continually. Scientists and engineers cope with such change and succeed in their careers by means of a continuous learning experience.

Learning throughout the career of a scientist or engineer involves three general criteria: (1) experience; (2) informal learning, such as reading journals and attending relevant technical meetings; and (3) and formal education and training programs. Through experience and education, the scientist and engineer will expand his/her general knowledge through formal means, while at the same time being trained by acquiring the specific skills required for a defined job function. Together the two comprise continuing education, the periodic career-long process that follows the degree-granting education of the scientist and engineer.

## 4.5   EDUCATION AND SPORTS PROGRAMS

There is currently considerable debate about the role and the effects of sports programs on education—to some extent in high schools but mainly in universities (Clotfelter, 2011). There is a fear that the assignment of money to such programs will cause a deficiency in the quality of education and, therefore, lead to gaps in the education and in the knowledge of the more scholastically inclined students at high-school level and particularly at university level. In fact, it is feared that at some universities, the sports programs fair better (in terms of financial support) than the majority of the education programs.

The issue of money assigned to sports versus money for education is an often debated topic in many educational institutions and in various newspapers. The focus is on the money spent on football, coaches, television contracts, and stadiums and there is a consistent worry about an imbalance between the expense of university sports programs and the challenge of funding academic learning enterprises (Clotfelter, 2011; Berkowitz, 2014).

Most of the statements to define or identity the amounts of money spent on sports criticize the opponents of sports programs by stating that they use extreme examples: spectacularly paid coaches of whom there is perhaps only a dozen or so out of the hundreds of college sports personnel, super-sized stadiums, and the existence of sports-department budgets when most sports programs operate on a more modest scale. The targets are attractive because the celebrity status of big-time football and basketball (1) fill pages of newspapers and specialty magazines, (2) appear endlessly on multiple television channels, and (3) enjoy the attention of rabid fans.

The defenders of university sports programs note that sports is a complicated enterprise that serves many interests at educational institutions—public and private universities, large and small. Sports in the universities are a pervasive part (some might say an invasive part) of the American education system, and like other high-profile activities (such as finance, real estate, or banking), there are bad actors, people of questionable integrity, and errors of commission and omission that attract justifiable or unjustifiable outrage and response (Bretag, 2013). However, neither of these views can offer support for nor detract from university sports programs. It is possible to see the advantages of the competitive world of university from a better perspective if intercollegiate athletics is considered in its various parts, including the engagement of students, the lives of student-athletes (both celebrity performers and regular participants), the involvement of alumni and public, and the financial consequences of sustaining these programs.

Money (and its assignment to various programs) in universities is always important, especially if the claim that very few universities subsidize athletics from student fees and general university revenue is true. Sports expenses are funded from earned revenue (tickets, television, sales, gifts, and similar revenue generated by the athletic activity itself), and from institutional revenue available for any purpose (student fees and university funds). The institutional revenue is a subsidy for an enterprise that, in the best of all possible worlds, should earn its own way in much the same fashion as other university nonacademic enterprises, such as food services, bookstores, parking, and housing.

On the other hand, using the university library as the example, the library is (or should be) a stable, standard, and continuing enterprise central to the meeting the needs of the educational programs of the university. The libraries in universities vary in size and complexity and are maintained to support instructional and research programs, compete for the best students and faculty, and also compete for the external funding that makes research at this level possible. Universities also require strong libraries for their success. The size of the libraries reflects an institutional commitment to the academic enterprise, while the subsidy for the sports program reflects a commitment to the nonacademic competitiveness of athletics. The subsidy also represents an institutional investment that the educational institution could have allocated to academic educational enterprises but instead uses to pay part of the cost of the intercollegiate athletic program, a nonacademic (even a noneducational) enterprise.

The impact of college sports on the academic enterprise also varies widely, from those institutions whose sports programs require no subsidy (and therefore have no detrimental impact on the academic enterprise) to those sports programs where the subsidy may be as much as the total library budget (Clotfelter, 2011). The impact of such a sports subsidy is an indication that sports at that institution do not compete well enough to earn sufficient revenue from attendance, television, sponsorships, alumni, and donors, and must spend university money to stay within the competitive context of the sports division in which the university team is placed. While talk of curtailing expenditures on sports is common and enthusiastic among many faculty and some outside commentators, the constituencies for college sports among alumni, trustees, elected officials, and fans are passionate at unbelievable levels. Trustees, alumni, and elected officials, in addition to fans of all kinds, want their sports regardless of the subsidy required at the expense of the academic enterprise.

Finally, to support the sports activity most universities have high-quality sports facilities—and these facilities are being continually improved with the passage of time. There will almost certainly be a sports center with at least a sports hall and other dry sports facilities including a fitness gym with modern cardiovascular and resistance machines, one or more exercise studios, and facilities such as squash courts. Many universities also have a pool and specialist facilities such as climbing walls. Outdoors, there are grassed fields (pitches) for different sports—generally of a very high quality—at least one floodlit artificial turf field (pitch), and possibly an athletics track and/or a boat house (if rowing is a university sport). Some universities also own or have access to outdoor centers for activities such as water sports, climbing, and hill walking. Very few (but some do) have their own golf course, but many have come to an arrangement with one or more local clubs, which allow students to use their courses at a reduced charge.

There are two sides to every coin (no monetary pun intended)—on the side of those in favor of university sports, the advocates note that winning teams stimulate private donations to the successful schools—often the donations might be listed in terms of the dollar amounts but the manner in which the money is to be used is not always disclosed. Systematic empirical evidence generally supports this, although the educational effects of the donations appear to be small, and result primarily from the appearance of football teams in post-season bowl games. There are similar stories of individual universities attracting more applications following athletic achievements, although in this case, the empirical evidence is questionable and there appears to be little effect on the academic credentials of classes enrolled subsequent to the athletic achievements. Although there has been much less attention focused on the effects of intercollegiate sports on the athletes themselves and other students, there appears to be a considerable number of conflicting implications for the intellectual atmosphere and achievements of university students from adopting a big-time sports program and the attendant culture.

If athletic success does boost donations and attracts more and better credentialed applicants to the successful institutions, there questions must be answered as to the origin of the students and whether or not the allocation of the resources is efficient and equitable. Until such concerns are addressed, it is impossible to decide if the indirect effects of college athletics are desirable or undesirable by looking at just one side of a reallocation of resources.

On the positive side (there really is a positive side), sports programs provide a range of benefits to campus life, such as exercise for students as well as publicity for the university. In addition, universities take pride in knowing that their sports departments help boost the local economy. However, in reality, to many observers, the net social welfare and equity implications of any indirect effects of college sports on institutions that host big-time intercollegiate teams really remain unknown. Issues such as (1) the direct or indirect economic value of intercollegiate athletic competition and (2) the cost of maintaining the football stadium, which seems to be one of the largest (if not the largest) structure on campus, as well as (3) the cost of maintaining all sports facilities during the off-season(s) remain to be answered satisfactorily. However, it would be interesting to note the number of the members of any university board of trustees (or board of regents) who would be willing to serve (on the board) if it was not for meetings being called at a time for the members to be treated to a freebee weekend with attendance at the relevant game accompanied by food and drink provided in the president's box or private lounge in the stadium. This is not good governance in any form whatsoever; in fact it is bad governance (Salmi and Helms, 2013).

In conclusion, there are many opinions that are for and against the prominent position of sports in university education systems. While empirical evidence may appear to support the concept of university sports, the effects appear to be small, and result primarily from the appearance of, say, football teams in postseason bowl games or the appearance of basketball teams in the so-called "March Madness." However, there appears to be little effect on the academic credentials of classes enrolled subsequent to such athletic achievements. Although there has been much less attention focused on the effects of intercollegiate sports on the athletes themselves and other students, there appear to be a number of conflicting implications for the intellectual atmosphere and achievements of university students from adopting a big-time sports program and its attendant culture (Getz and Siegfried, 2010).

Attempting to determine whether or not the effects of college athletics are desirable or undesirable by looking at just one side of a reallocation of resources is not the correct method of evaluating sports programs, as the results can be biased to favor one side or the other of the discussion. In fact, the net benefit of any direct or indirect effects of college sports on institutions that host big-time intercollegiate teams really remains unknown. It is possible that these effects could be sufficiently large and undesirable to outweigh the any surplus created by the direct entertainment value of intercollegiate athletic competition (Getz and Siegfried, 2010).

## 4.6   TEACHING ETHICS

Scientific and engineering disciplines are considered to be highly ethical professions in which the scientists and engineers exhibit behavior that is of the highest ethical and moral standards. Ethics is "the normative science of conduct, and conduct is a collective name for voluntary actions" (Lillie, 2001, p. 3). In this regard, voluntary actions are those actions that could have been done differently and such actions may be good or bad, right or wrong, moral or immoral. Ethics focuses not on what scientists and engineers think, but what they ought to think and do (Lillie, 2001; Howard and Korver, 2008).

Whatever the definition, ethics is one of the pillars of scientific and engineering research. It is definitely one of the criteria for evaluating the quality of higher education in these aforementioned areas. Despite the range of factors that contribute to ethical or unethical behavior, the central determinants are the personal thoughts (and behavior) of the scientist and engineer, which determine the meaning that an individual attaches to his/her position with respect to ethics. Furthermore, the ethical aspects of scientific and engineering research revolve around the responses to (1) the ethically proper way to collect, analyze, and report all aspects of a study, and (2) the researcher–respondents interaction, which is especially true in the social sciences where surveys of human actions and interactions are accumulated and interpreted (Kitchener and Kitchener, 2009, p. 6).

The gaps in knowledge that arise from incorrect and improper teaching in schools and universities are tangible results that can be determined and corrected. However, the biggest knowledge gap that schools and universities must address is the teaching of ethics as part of science and engineering courses, and this is not always a tangible issue.

Nevertheless, ethics is one of the pillars of higher education in scientific and engineering, in terms of research, teaching, and community service requirements. (Whitbeck, 1995; Speight and Foote, 2011). It is definitely one of the criteria for evaluating the quality of higher education in these aforementioned areas. Despite the range of factors that contribute to ethical or unethical behavior, the central determinants are the personal thoughts (and behavior) of the scientist and engineer, which determine the meaning that an individual attaches to his/her position with respect to ethics.

Many scientists and engineers have gaps in knowledge in pertaining to correct behavior and professionalism. In other words, many scientists and engineers do not understand nor are they taught the concept of ethics, and hence they may not feel bound by any standards of behavior when dealing with others as members of a team or as individuals. Because of this, unethical behavior can and does arise and may be persistent (until caught) in many scientists and engineers (Whitbeck, 1995; Speight and Foote, 2011). Thus it is pertinent that space be assigned here for a discussion of the need for ethics to be taught at schools and, if necessary, at institutes of higher education as part of the education of scientists, engineers, and other professions.

To bridge the gap in knowledge in terms of ethical behavior, teaching morals and ethical values to the students should begin at home! In the education system it begins in schools, where unfortunately cheating is not unknown (Speight and Foote, 2011; Anderson and Kamata, 2013; Bretag, 2013; Essoyan, 2014). If the tendency for students to cheat is not curbed, the concept of cheating becomes ingrained in the students' psyche as a natural phenomenon and continues at university and thence unto adult life. In such situations, the students who are colleagues of the cheater must not fail to report cheating in any form (Cheung, 2014).

This is well illustrated by the movie *The Emperor's Club* (starring Kevin Kline, Emile Hirsch, and Joel Gretsch; directed by Michael Hoffman; written by Ethan Canin and Neil Tolkin; distributed by Universal Pictures), in which an idealistic prep-school teacher at a fictitious private school attempts to redeem an incorrigible student who cheats to win a top place in a competition, being successful in this nefarious activity, the tendency to cheat continues into the adult life of the student.

The movie illustrates an excellent example of the need for teachers/professors at universities to promote values within science and engineering fields that fit the needs of modern industries. The efforts of developing countries to achieve developed status, with a focus on science and engineering, and the initiation of industries along with other economic and political institutions, has opened the doors for new values and challenges in the field of science and engineering. It is essential that university curricula examine these challenges and educate scientists and engineers to confront and present solutions for them.

Thus, a main objective in promoting morals and values education for scientists and engineers is to encourage universities to implement academic and other activities related to teaching, research, and extension programs embracing values and culture, such as seminars, conferences, workshops, and orientation programs for both science and engineering lecturers and their students. In addition, universities can also produce materials related to morals and values education.

Furthermore, the scientific and engineering communities are typically divided into two main groups: (1) those who work with a good and clear conscience to produce results that will help mankind and (2) those who work for their own glory and betray the respective science and engineering disciplines through unethical behavior, which can involve stealing the ideas of others or changing experimental data to prove their own hypotheses (Speight and Foote, 2011). This is where mentoring can play a major role in maintaining honesty in scientific and engineering professions. Scientists and engineers who fall into the first category must be willing to take on the role of watchdog in making sure that such unethical activities are stopped at the starting gate and not allowed to move any further, even if it means reporting on and possibly resulting in the termination of the career(s) of the miscreant(s).

The concept of duty is thus also a necessary part of the education of scientists and engineers. The scientist or engineer must never think to herself/himself that s/he is only responsible for generating scientific or engineering knowledge.

*Truth* in science and engineering ensures that the scientist or engineer has a vision that arises from his/her education and gives the direction to the next step. Erroneous or faulty knowledge or knowledge gaps (Chapter 4) mars the thinking process and the next step does not proceed in the right direction. That step might allow the scientist or engineer to develop new power and be looked upon with awe by others who have not advanced as far in their respective careers. But it cannot be certain that the progress will be responsible, or available for the benefit of mankind. Such problems exist because of the lack of understanding of faulty data (obtained by unethical practices) (Speight and Foote, 2011). It is necessary that the scientist or engineer be sufficiently well educated to the extent that s/he can recognize such practices.

Scientists and engineers use the knowledge gained by honest and careful experimentation to decide the means by which to apply that knowledge. If there is fault in the data, education should assist the professional to recognize that the application of faulty data may have destructive consequences. Science or engineering knowledge is often considered to be value-neutral insofar as there *should* be no bias in the experimental procedures by which the knowledge was obtained. But it is necessary to discover and retain the scientific and engineering spirit, which is always more important than the technique, the knowledge, or the method in any scientific or engineering activity.

In reality, the spirit of science and engineering is based on humility—this commences with the admission that not all of the truth about nature and natural events is known—and the education system should teach scientists and engineers to form theories and look (without bias) for methods by which they can test whether or not a theory is correct, as well as to what extent that theory is correct. The scientist or engineer who has been taught correctly should not ignore data that does not support the theory, nor should they change the data (Speight and Foote, 2011). It is only by following the truth that science and engineering will continue to progress. The education process should have instilled into the students that nothing should be accepted on the basis of the authority and position (or beliefs) of the lead investigator. The education process teaches that science and engineering demand proof obtained through (1) observation, (2) testing with experiments, and (3) truth, which must be universal and which everybody can be assured is the truth.

## 4.7  THE GENDER GAP

While some purists may consider (or may even argue) this section should not be included in a chapter dealing with gaps in knowledge, it is very relevant and necessary to include such material. Acknowledgment of women in science and engineering is very necessary and not to do so represents a gap in knowledge. Women form an important part of the scientific and engineering communities and bring to the table many issues that might not be foreseen by their male counterparts, and they often pay attention to details that a male scientist or engineer might pass over as being insignificant. Women must be given at least equal status to male scientists and engineers.

In fact, as a case in point, Marie-Anne Pierrette Paulze (Madame Lavoisier, as she was more commonly known) dutifully sat and took copious notes as her husband, Monsieur Antoine Lavoisier—the famed seventeenth-century chemist—worked in the laboratory (Poirier, 2012). Not many people are aware of the work of Madame Lavoisier or even bother to consider her diligence in recording her husband's work. In fact, had she not been familiar with the chemical aspects of her husband's work and had an understanding of the scientific aspects of his work, the discoveries made by Monsieur Lavoisier would have been lost to history when the guillotine brought an end to their lives in 1794. Perhaps the guillotine robbed her of due recognition— one will never know the truth of the matter.

Another case is that of Dr. Rosalind Franklin, who made critical contributions to the understanding of the fine molecular structures of DNA (deoxyribonucleic acid) and RNA (ribonucleic acid) viruses, along with other carbonaceous materials such as coal and graphite (Franklin, 1946–1959; Olby, 2012). She was the first to observe the first X-ray diffraction patterns of the DNA molecule, which led to the derivation of the double helix structure, although this was ignored when it came to handing out the Nobel Prize kudos. This is considered by many observers, and justifiably so, to be a travesty and a major omission in the recognition of a woman in science. Sadly, Dr. Franklin died in 1958 of ovarian cancer at the age of 37 and was never able to receive full appreciation for her work.

In fact, going along with this theme, women were for many decades not considered to be major contributors in scientific and engineering research. Women were tagged on to the authorship of published manuscripts more as helpers rather than as major contributors to the work. The occasional woman made it into the so-called top tier of scientists or engineers, but she was looked upon as an aberration rather than a person with knowledge or talent.

The modern female scientist or engineer believes that the issues of concern for women have been addressed, and that women are now incorporated into the mainstream of scientific and engineering education and research. Yet in many countries participation rates for women in scientific and engineering research remain low and recognition is not forthcoming. In fact, looking back more than 50 years of involvement in scientific and engineering research, there are certainly more women in science and engineering than there used to be in the 1950s, 1960s, 1970s, 1980s, 1990s, and 2000s (NRC, 2006). It is, of course, pertinent to determine if this trend has continued or, as it may appear, the numbers have stagnated and perhaps even decreased. Alternatively, there may be a preference for one scientific and engineering discipline over another.

The participation of women in science and engineering programs has suffered because of gender stereotypes in labor division, which have defined many science subcategories and engineering as male-oriented professions (Cockburn, 1985; McIlwee and Robinson, 1992; Kvande and Rasmussen, 1994; Gherardi, 1995; Sonnert and Holton, 1995; Frehill, 1997; Kvande, 1999; Faulkner, 2000; Hersh, 2000). In addition, the traditional university-based learning environment in science and engineering programs (with some gender bias by lecturers/professors) has favored male student interest and ignored the preferences of female students (Hacker, 1989; Tonso, 1996; Seymour and Hewitt, 1997; Salminen-Karlsson, 2002).

These gender biases have influenced the educational choices of male and female students, leading to a disproportional distribution of male students in the physical sciences, natural sciences, mathematics, and engineering and a disproportional distribution of female students in arts, humanities, and social sciences (McIlwee and Robinson, 1992; Lackland and De Lisi, 2001; Smith, 2005). A common perception that there are many of women in certain fields of science and engineering, such as the environmental and biological subdisciplines, may be true. Such disciplines did not harbor the so-called cold shoulder or chilly climate that women still experience in other subdisciplines, with often blatant discrimination not only from fellow (male) students but also from the faculty (Conefrey, 2001; Leach, 2013).

Many of the science and engineering programs with a high percentage of women have been regarded as having less status than the male-dominated traditional programs—they are regarded as "soft." Any entirely new program created to recruit women, it was believed, should have "soft" features, but to stress these soft features would place it far down in the hierarchy. It was feared that, if any new program were to have too feminine an image, it would interest neither male nor female students. This created a twofold dilemma: any program created to attract female students may be unattractive to female students (Salminen-Karlsson, 2002). To effect change in a prevailing curriculum, the basic unit (a university department) is perhaps the most important aspect (Becher and Kogan, 1992).

The perception is that women are sociable and like working in groups—to further such a premise the concept of problem-based learning (which still receives mixed reviews) requires that study groups are a fundamental learning device. There is the idea that male students might be prone to be dominant in mixed groups, while in reality, it is the female students who do most of the work in groups, while many of the male participants coast along and get the benefits of their hard-working female counterparts.

As long as the system and ideology behind education in science and engineering remains static, the stereotyped opinion that women in science and engineering education are different to other women will remain firmly rooted in the psyche of teachers and professors, along with the idea that female students should remain subservient to male students.

## REFERENCES

Anderson, M.S. and Kamata, T. 2013. Scientific research integrity as a matter of transparency. In: *Global Corruption Report: Education* (Chapter 3.19). Routledge, Taylor & Francis Group, Abingdon, Oxford.

Balf, T. 2014. The SAT is not fair. *New York Times Magazine*. Sunday, March 9, pp. 26–51.

Becher, T. and Kogan, M. 1992. *Process and Structure in Higher Education*. Routledge, New York.

Berkowitz, S. 2014. $5M Pay, title experience not needed. *USA Today*. January 14, p. 1A.

Bettinger, E.P., Long, B.T., Oreopoulos, P., and Sanbonmatsu, L. 2009. The role of simplification and information in college decisions: Results from the H&R block FAFSA experiment. Working Paper No. 15361, NBER Working Paper Series, National Bureau of Economic Research, Cambridge, MA.

Bransford, J.D., Brown, A.L., and Cocking, R. (eds). 2000. *How People Learn: Brain, Mind, Experience, and School*. National Research Council, National Academy Press, Washington, DC.

Bretag, T. 2013. Short-cut students: From academic misconduct to academic integrity. In: *Global Corruption Report: Education* (Chapter 3.13). Routledge, Taylor & Francis Group, Abingdon, Oxford.

Buitelaar, P. and Cimiano, P. (eds). 2008. *Ontology Learning and Population: Bridging the Gap Between Text and Knowledge*. IOS Press, Amsterdam, The Netherlands.

Cheung, J. 2014. The fading honor code: If students fail to report cheaters, can a culture of integrity prevail. *Education Life, New York Times*. April 13, pp. 25–26.

Clotfelter, C.D. 2011. *Big Time Sports in American Universities*. Cambridge University Press, Cambridge, United Kingdom.

Cockburn, C. 1985. The material of male power. In: M. Donald and J. Wajcman (eds), *The Social Shaping of Technology*, pp. 125–146. Open University Press, Milton Keynes, United Kingdom.

Conefrey, T. 2001. Sexual discrimination and women's retention rates in science and engineering programs. *Feminist Teacher*, 13(3): 170–192.

Elmore, R.F. 2002. *Bridging the Gap Between Standards and Achievement: The Imperative for Professional Development in Education*. Albert Shanker Institute, Washington, DC.

Engberg, M.E. and Wolniak, G.C. 2010. Examining the effects of high school contexts on post-secondary enrollment. *Research in Higher Education*, 51: 132–153.

Essoyan, S. 2014. *Cheating on Tests Spurs KCC to Change Exam Rules*, pp. A1–A8. Star Advertiser, Honolulu, HI.

Faulkner, W. 2000. Hierarchies and gender in engineering. *Social Studies of Science*, 30(5): 759–792.

Franklin, R.E. 1946–1959. Selected papers and the Rosalind Franklin papers: Biographical information. http://en.wikipedia.org/wiki/Rosalind_Franklin (accessed March-August 2013); http://profiles.nlm.nih.gov/ps/retrieve/Narrative/KR/p-nid/183 (accessed June 30, 2013).

Frehill, L.M. 1997. Education and occupational sex segregation: The decision to major in engineering. *Sociological Quarterly*, 38(2): 225–249.

Getz, M. and Siegfried, J.J. 2010. What does intercollegiate athletics do to or for colleges and universities? Paper No. 1005. Working Papers, Vanderbilt University Department of Economics, Vanderbilt University, Nashville, TN. http://www.accessecon.com/pubs/VUECON/vu10-w05.pdf Revised version, May 2010 (application/pdf); (accessed July 19, 2013).

Gherardi, S. 1995. *Gender, Symbolism and Organizational Culture*. Sage Publications, Thousand Oaks, CA.

Ginsberg, S.M. 2010. Mind the gap in the classroom. *The Journal of Effective Teaching*, 10(2): 74–80.

Glencorse, B. 2013. Testing new tools for accountability in higher education. In: *Global Corruption Report: Education* (Chapter 4.14). Routledge, Taylor & Francis Group, Abingdon, Oxford.

Gorcham, S. 2013. *The Mad, Mad, Mad World of Climatism: Mankind and Climate Change Mania*. New Lenox Books, New Lenox, IL.

Grodsky, E. and Jones, M.T. 2007. Real and imagined barriers to college entry: Perceptions of cost. *Social Science Research*, 36(2): 745–766.

Hacker, S. 1989. *Pleasure, Power, and Technology: Some Tales of Gender, Engineering and the Cooperative Workplace*. Unwin Hyman, Boston, MA.

Hersh, M. 2000. The changing position of women in engineering worldwide. *IEEE Transactions of Engineering Management*, 47(3): 345–359.

Hill, L.H. 2008. Concept mapping to encourage meaningful student learning in adult education. *Adult Learning*, 16(3/4): 7–13.

Horn, L.J. and Chen, X. 1998. Toward eesiliency: At-risk students who make it to college. Report No. PLLI-98-8056. United States Government Printing Office, Washington, DC.

Horn, L.J., Chen, X., and Chapman, C. 2003. Getting ready to pay for college: What students and their parents know about the cost of college tuition and what they are doing to find out. Statistical Analysis Report No. NCES 2003-030. National Center for Education Statistics, United States Department of Education, Washington DC.

Howard, R.A. and Korver, C.D. 2008. *Ethics for the Real World*. Harvard Business Press, Boston, MA.

Ingels, S.J. and Dalton, B.W. 2008. Trends among high school seniors, 1972–2004. Report No. NCES 2008-320. National Center for Education Statistics, Institute for Education Sciences, United States Department of Education, Washington, DC.

Kitchener, K.S. and Kitchener, R.F. 2009. Social science research ethics historical and philosophical issues. In: D.M. Mertens and P.E. Ginsberg (eds), *The Handbook of Social Research Ethics*. Sage Publications, Thousand Oaks, CA.

Kvande, E. 1999. In the belly of the beast—Constructing femininities in engineering organizations. *The European Journal of Women's Studies*, 6: 305–328.

Kvande, E. and Rasmussen, B. 1994. Men in male-dominated organizations and their encounter with women intruders. *Scandinavian Journal of Management*, 10(2): 164–175.

Lackland, A.C. and De Lisi, R. 2001. Students' choices of college majors that are gender traditional and nontraditional. *Journal of College Student Development*, 42(1): 39–48.

Leach, F. 2013. Corruption as abuse of power. In: *Global Corruption Report: Education* (Chapter 2.13). Routledge, Taylor & Francis Group, Abingdon, Oxford.

Lillie, W. 2001. *An Introduction to Ethics*. Allied Publishers Limited, New Delhi, India.

McIlwee, S. and Robinson, G. 1992. *Women in Engineering: Gender, Power, and Workplace Culture*. State University of New York Press, Albany, NY.

Neuman, S.B. and Celano, D. 2006. The knowledge gap: Implications of leveling the playing field for low-income and middle-income children. *Reading Research Quarterly*, 41(2): 176–201.

NRC (National Research Council). 1985. *Continuing Education of Engineers: Engineering Education and Practice in the United States*. Panel on Continuing Education Committee on the Education and Utilization of the Engineer, Commission on Engineering and Technical Systems, National Research Council, National Academy Press, Washington, DC.

NRC (National Research Council). 2006. *To Recruit and Advance: Women Students and Faculty in U.S. Science and Engineering*. Committee on the Guide to Recruiting and Advancing Women Scientists and Engineers in Academia. Committee on Women in Science and Engineering Policy and Global Affairs, National Research Council of the National Academies, National Academies Press, Washington, DC.

Olby, R. 2012. Francis Crick and James Watson: Decoding the structure of DNA and the secret of life. In: A. Robinson (ed.), *The Scientists: An Epic of Discovery*. Thames & Hudson, London.

Pérez-Peña, R. 2014. What you don't know about financial aid (but should). *Education Life, New York Times*. April 13, pp. 18–21.

Plank, S.B. and Jordan, W.J. 2001. Effects of information, guidance, and actions on post-secondary destinations: A study of talent loss. *American Educational Research Journal*, 38(4): 947–979.

Poirier, J.P. 2012. Antoine-Laurent de Lavoisier. In: A. Robinson (ed.), *The Scientists: An Epic of Discovery*. Thames & Hudson, London.

Salmi, J. and Helms, R.M. 2013. Governance instruments to combat corruption in high education. In: *Global Corruption Report: Education* (Chapter 3.2). Routledge, Taylor & Francis Group, Abingdon, Oxford.

Salminen-Karlsson, M. 2002. Gender-inclusive computer engineering education: Two attempts at curriculum change. *International Journal of Engineering Education*, 18(4): 430–437.

Seymour, E. and Hewitt, N. 1997. *Talking About Leaving—Why Undergraduates Leave the Sciences*. Westview Press, Boulder, CO.

Smith, E. 2005. Gender differentiation and early labor market integration across Europe. *European Societies*, 7(3): 451–479.

Sonnert, G. and Holton, G. 1995. *Gender Differences in Science Careers*. Rutgers University Press, New Brunswick, NJ.

Speight, J.G. and Foote, R. 2011. *Ethics in Science and Engineering*. Scrivener, Salem, MA.

Stensaker, B. 2013. Ensuring quality in quality assurance. In: *Global Corruption Report: Education* (Chapter 3.5). Routledge, Taylor & Francis Group, Abingdon, Oxford.

Stiglitz, J. 1999. Knowledge as a global public good. In: I. Kahl, I. Grunberg, and M.A. Stern (eds), *Global Public Goods: International Cooperation in the 21st Century*, p. 317. Oxford University Press, Oxford.

TEAC (Teacher Education Accreditation Council). 2014. http://www.teac.org/ (accessed June 1, 2014).

Tonso, L. 1996. The impact of cultural norms on women. *Journal of Engineering Education*, 86(3): 217–225.

Weimer, M. 2009. *Helping Students Fill Gaps in Basic Knowledge. Philosophy of Teaching*. Faculty Focus, Madison, WI, December 15.

Whitbeck, C. 1995. Teaching ethics to scientists and engineers: Moral agents and moral problems. *Science and Engineering Ethics*, 1(3): 299–308.

Yang, J. and Grabe, M.E. 2011. Knowledge acquisition gaps: A comparison of print versus online news sources. *New Media & Society*, 13(8): 1211–1227.

# 5 Bridging the Knowledge Gap

## 5.1 INTRODUCTION

The United States has experienced and continues to experience an almost unprecedented sharp increase in knowledge inequality (Neuman, 2006), sometimes referred to as a knowledge gap, (Tichenor et al., 1970; Donohue et al., 1975; Gaziano, 1997; Hwang and Jeong, 2009) which is evident not only in individual students but also between groups of students. In fact, the school–university knowledge gap in science and engineering has regularly been both acknowledged and denounced over the last 3–4 decades (Chapter 4). In addition, acceptable methods of addressing the knowledge gap have not been developed.

The process of teaching science and engineering (or mathematics) from meaningful texts, at least up to preuniversity-level schooling for all students, may be a way to remove the dilemma of the knowledge gap (Neuman and Celano, 2006). This approach to teaching science and engineering can act as a link connecting textbooks and school-level laboratory work to the real world of science and engineering. Furthermore, teachers/professors must nurture the scientific and engineering professions by teaching students methods of experimentation (with verification of the experimental data), observation, and deduction—such teaching can lead to more realistic professional development in the fields of science and engineering.

Many observers, who may not understand the concepts of science and engineering, have erroneously assumed or stated that there is no difference between a scientist and an engineer (much to the chagrin of the scientist and engineer), while other observers think (also erroneously) the two careers are totally separate from each other. A very general difference between the scientist and the engineer is that the scientist discovers new theories and uses them to solve problems scientifically, while engineers use the theories discovered by scientists to solve problems technologically—this idea may seem closer to the truth but also suffers from flaws in logic. However, it must be recognized that the roles can also be reversed, and, in the present context, the simple and convenient definitions of a scientist and engineer—that is: (1) a scientist is a person who has scientific training or who works in the sciences, while (2) an engineer is someone who is trained as an engineer—may be not quite true.

In the simplest sense, the practical difference between a scientist and engineer lies in the educational degree and the description of the task being performed by the scientist or engineer. On a more philosophical level, scientists tend to explore physical phenomena to discover new knowledge. Engineers take that knowledge and *apply* it to solve practical problems, often with a necessary bias (quite often not seen in the scientist) toward optimizing cost or efficiency. On a more realistic level, and not to short-change one or the other of these disciplines, scientists can develop processes and process equipment, while engineers can initiate and discover new concepts. In fact, there is (if not, there should be) considerable overlap between science and engineering—there are scientists who design and construct equipment and engineers who make important scientific discoveries.

The true issue is not so much a correct definition of a scientist or an engineer but the means by which a balance can be stuck between science and engineering for the benefit of both groups (and their respective subgroups) of professionals. Part of the answer lies in the education system in which scientists and engineers are introduced to the parameters of the various disciplines so that crossfertilization can occur between science and engineering. It is not to be inferred that scientists should be brow-beaten with engineering facts and vice versa. In this sense, a simple and enjoyable introduction is necessary between the two main disciplines so that each discipline (and the respective subdisciplines) can appreciate the basic knowledge taught in the other. A basic course for the scientist involving the principles of engineering will help that scientist formulate the right questions to be asked and to know where (or to whom) to turn to s/he can find the answer(s). In one way or another, both the scientist and engineer use their educational background to solve technical problems and their work cannot be differentiated without overlap—the obvious comparisons are between chemists and chemical engineers.

The purpose of this chapter is to present selected areas of overlap and how the overlap might be enhanced. The obvious method of enhancing the overlap is (1) to educate scientists in the basic principles of engineering and (2) to educate engineers in the basic principles of science. In both cases there should be a initial special focus on the areas where both disciplines overlap, after which the student can move on to more areas where the disciplines also overlap but which may be somewhat less than obvious to the neophyte student.

## 5.2   SCIENCE FOR ENGINEERS

Engineering in the modern sense traces its name back to the pre-Roman era. For example, engineers designed and built the Egyptian pyramids, the temples of pre-Christian Greece, the Great Wall of China, and the Hanging Gardens of Babylon (Chapter 1). Roman engineers built roads and aqueducts, and designed under-floor heating systems, drainage systems, and sewer systems, among other numerous accomplishments. However, there are other aspects of engineering that need to be addressed and that is the overall knowledge of the engineer. In the light of the accomplishments presented in the last paragraph, it might be asked that since engineers have built the world, who needs science? The answer is: read on.

Using chemistry as the example of the focus of the course, engineering requires applied science, and since chemistry has very applied applications the more chemistry an engineer understands, the more beneficial it is. In this sense, chemical engineering basically is often considered to be applied chemistry—not so. Chemical engineering is, in fact, the branch of engineering concerned with the design, construction, and operation of machines and plants that perform chemical reactions to solve practical problems or make useful products. In such a course, the concepts and theory of chemistry must be shown to be related using examples from fields of practical application, thus reinforcing the connection between science and engineering. The topics covered should provide the student with the fundamental tools necessary for use by an accomplished engineer.

In fact, there are many courses generally labeled as "science for engineers" or some similar title. Typically, the courses are intended to provide engineering students with a background in important concepts and principles of science. Nonengineering majors, including physicists and geophysicists, could also benefit from such a course. In the courses emphasis is placed on those areas considered most relevant in an engineering context, and practical applications in engineering, and technology may also be part of such a course. Indeed, it is the intent of such courses to give the student a deep understanding of the science (as included in the course content) and not just shallow memorization of seemingly unrelated and even distorted facts or equations.

Like all engineers from other subdisciplines of engineering, chemical engineers use mathematics, physics, and economics to solve technical problems. The difference between chemical engineers and other types of engineers is that they apply knowledge of chemistry in addition to other engineering disciplines. In addition, some chemical engineers make designs and invent new processes while others construct instruments and facilities and yet others plan and operate facilities. Chemical engineers have helped develop ideas in fields such as atomic science, polymers, paper, dyes, drugs, plastics, fertilizers, foods, and petrochemicals. They devise ways to make products from raw materials and ways to convert one material into another useful form. Chemical engineers can make processes more cost effective, or more environmentally friendly, or more efficient. Moving to the other end of the engineering spectrum, a civil engineer may not be seen (by some observers, including some teachers and professors) as needing science as a background. Yet, for the civil engineer involved in various building projects, a background knowledge of the chemistry of cement might be very useful—especially a knowledge of the interaction of the various components of cement and mortar, the nature of pozzolanic materials, and the means by which cement or mortar binds to other surfaces sufficiently well to bear the loads put upon the structure.

Continuing with chemistry for the moment (not because it is the *chosen* science but merely used as an example here), chemical principles are also involved in science and engineering disciplines such as environmental science and environmental engineering. As an example, whenever an oil spill occurs on water (whether the sea or land-based bodies of water, such as a lake or river) the goal is to clean up *all* of the oil. The engineer needs to know that in most cases the oil oxidizes (due to aerial oxygen) and that the oxidation process causes a change in the density of the oxidized constituents of the oil. As the change in density occurs, the oil passes from being lighter than water to being possibly the same density as water or even heavier than water. Thus, part of the oil (the oxidized constituents) disappear from the surface and end up floating below the water surface or even sinking to the bottom. This can lead to a false indication of the *total* recovery of the oil when, in actuality, part of the oil is under the surface of the water or on the bottom of the sea, lake, or river. Through the agency of other chemical reactions, the ignored constituents of the oil will reappear at a later date (days, months, or even years after the spill) and continue to cause damage to what was thought to be a clean environment. And this is not a fanciful scenario; it has happened before and will continue to happen as long as the chemistry of oil oxidation is ignored by the cleanup team, who may be mostly engineers with a very minimal, if any, background in chemistry.

A similar argument can be made for the influence of bacteria on an oil spill. The bacteria consume the hydrocarbon constituents (the lower density molecules) of the oil, leaving the more complex (and higher density) molecules to disperse in the environment. As above, due to the occurrence of other biochemical or chemical reactions, the ignored oil will reappear at a later date and continue to cause damage to what was thought to be a clean environment. This is not to suggest that all engineers suffer through the same course details as chemists, biochemists, or environmental chemists. It does, however, suggest that an engineer with a smattering of chemistry (or biochemistry) will have the wherewithal to realize that all is not well with, in this case, oil cleanup, and will have the basic knowledge to know when and at what stage of such a project (preferably at the beginning) to call in experts in other fields.

The challenge in overlapping the fields of science and engineering education where and when necessary is to prevent the frequent disconnect between the scientist and engineer when a common discourse could be very fruitful. For example, say a chemist is trying to understand laboratory work from a fundamental mechanistic view, while an engineer needs some numbers (data) to plug into the relevant engineering (process-related) equations. There is a tendency for the scientist, when asked for help, to work on further experiments to generate the data the engineers need. Worthy as this may seem, the problem is compounded by the fact that many engineers (because of a knowledge gap) lack the practical skills to generate the data themselves. The conversation may then be terminated (wrongly and without justifiable cause) by the scientist, who may state that s/he does not have the time since s/he is fully engaged in the next project.

## 5.3   ENGINEERING FOR SCIENTISTS

In the simplest sense, engineers address real-world, out-of-the-laboratory problems. They improve ideas that have been demonstrated to be scientifically viable, and during the course of their work new concepts may also be initiated and developed as the engineer applies principles of education to try to find the most logical or most appropriate answer to a question.

Using chemical engineers and an example, some chemical engineers design process equipment and process flow sheets—as an example, such design is crucial in the refining industry where a refinery is a collection of unit processes that must operate in unison or the refinery shuts down. Thus, chemical engineers (not forgetting contributions by metallurgical and structural engineers) can make processes more cost effective, more environmentally friendly, or more efficient. To the scientists (in this case, the chemist) this may not mean much since s/he (the scientist) focuses on work in the laboratory. Nevertheless, there is the ever-present need for the scientist to remember that any concept conceived and demonstrated in the chemical laboratory has to go through a thorough engineering examination (pilot plant) and demonstration (demonstration plant) before it can even be considered to be placed on line in the real world of commercial processes.

Engineering courses for scientists need to assist and show the scientist the means of bridging the gap between the research lab and the industrial process. Anyone involved in research and development, quality control, design, development, production, or processes needing an overview of modern chemical engineering will benefit from such a course. The challenge is to assure that the practicing chemist understands what is needed to take a chemical concept to industrial process.

For example, once a concept has been shown to be workable in the laboratory, other issues arise such as (1) the ready availability of reactants or feedstocks; (2) safety and handling of the reactants or feedstocks; (3) process fundamentals, such as efficient conversion to products and the yield of the products; (4) process fundamentals, such as material and energy balance; (5) reactor design, reaction systems, and process economics; (6) process details, such as thermodynamics, reactor design, and transport phenomena; and (7) process integration, that is, process control and design. The final list may not necessarily be in this order, but all of these items (as well as others not listed here) need to be taken into consideration. Work in the chemical laboratory will cover some of the items, but not all.

It is of benefit if the concept-initiating laboratory scientist can consider each of these categories, and this is where an understanding of the basic principles of engineering will be advantageous. This will have given him/her (the concept-initiating laboratory scientist) the means to apply (some) chemical-engineering knowledge to further the bench-scale process and understand the important chemical engineering principles involved in the further development of the concept. Such knowledge is useful for those scientists who need to negotiate and interact with engineers and who need to appraise the industrial feasibility of their newly conceived chemical process.

In this context, the laboratory chemist will understand the ease or difficulty of choosing a replacement compound for a particular application. This individual will also understand that small changes in the structures of organic compounds can have a large effect on reaction rate and product properties. While the engineer may not be expected to be skilled in such details, the mere knowledge that such differences can affect the process (and the various process parameters as well as reactor design) will give the engineer the impetus to seek expert opinions—typically the inventor of the concept or a person with detailed chemical skills in this matter. Chemists often intuitively know some (but not all) of these effects, but may not understand everything about what gives a chemical compound properties that are specific to that compound.

Although the examples presented here have focused on chemists and chemical engineers as their work relates to the initiation and development of commercial processes, for whatever the reason many scientists are ready and willing to collaborate with engineers. Some of the major challenges facing the university education system include the means by which such interactions can be fostered. Shared education and appreciation between the various disciplines is vital, and it is to be hoped that universities will recognize this necessity. The old adage "never the twain shall meet"—referring in this context to scientists and engineers—is not acceptable.

## 5.4   THE MISSING LINKS

During the past several decades, there have been numerous studies conducted by organizations such as the national academies, federal agencies, business organizations, and professional societies suggesting the need for new paradigms in science and engineering practice, research, and education that better address the needs of a twenty-first-century nation (Avenier and Bartunek, 2010). However, despite the ever-growing importance of science and engineering practice to the future, these professions still tend to be held in relatively low regard compared to other learned professions such as the legal and medical professions.

There is still the tendency (by corporations and the public) to view scientists and engineers as disposable commodities that are to be discarded when the skills (or hands-on methods) become obsolete or replaceable by cheaper science and engineering services from other countries. In the case of corporations, when the bottom line is not financially healthy, it is the scientists and engineers who are dismissed/discarded/laid off rather than the corporate managers who may have overestimated the potential for revenue or mismanaged their respective division. At the same time, the bottom line notwithstanding, the chief executive officer may receive his/her multimillion-dollar salary plus bonus, or, if is s/he is responsible for shipping jobs to foreign countries (and other unacceptable actions such as driving the stock in a downward spiral to the lowest levels in recent company history) to the dismay of the board of directors, s/he will receive a multimillion-dollar severance package. Had the board been skilled in governance (as each member should be) (Salmi and Helms, 2013), such pay-offs could be mitigated. The scientists and engineers may be written up as being out of touch and suffering a knowledge gap that will not allow the company to remain competitive in the dog-eat-dog world of corporate technology and technology management.

As a result, and in the interests of further cost cutting to save the floundering bottom line and stock price, the company may also outsource an even greater number of scientific and engineering projects to other countries on the basis that it not only saves the company money but also keeps the company technically up-to-date. The former reason (i.e., saving money) may be (and often is) the real reason for the outsourcing of scientific and engineering projects. The second reason (i.e., keeping the company technically up-to-date) is often pure fiction and is no real reason to disembowel the company of real scientific and engineering talent and at the same time remove from the financial liability of the company the higher paid older scientists and engineers. These older scientists and engineers are the men and women who can save the company millions of dollars by recalling from memory which past scientific or engineering projects worked, which (although justifiable) did not, and which (i.e., perhaps the preferred project of some manager or other) were a complete waste of time and money. Typically, the company administration and finance office refuse to acknowledge such benefits of maintaining an experienced work force. Their interest is purely—but not always with complete justification—the bottom line.

At such time, the howls coming from various (unknowledgeable) mangers remind everyone that the "old boys" (and the "old girls") are not up-to-date in their thinking and are living in the past. Without saying it in as many words, the (unknowledgeable) mangers have latched on to the catch-phrase "knowledge gap" and their only plan to remove the knowledge gap is to discharge (lay off) the scientists and engineers with knowledge and recruit younger individuals or scientists and engineers from other countries who are devoid of knowledge.

While the companies then bask in the no-longer-red-but-black glow of the bottom line (if the company bottom line was ever in the red) the schools and universities maintain the standard of teaching that have placed science and engineering graduates to the current stage of a where there remains a large and obvious knowledge gap. Indeed, to a large degree, the differences between scientists and engineers are driven by the current education system. In terms of the deficiencies of the scientist, there is some truth in the charge that s/he will not think quantitatively but is quite happy when a concept has been proven, although the reaction parameters are not fully understood, and s/he is unable to speak the same language as the engineer (Houlton, 2013). On the other hand, in terms of the deficiencies of the engineer, there is also some truth in the charge that s/he does not understand the general scientific principles behind the process and is unable to speak the same language as the scientist.

Science (typically chemistry) at school and at university tends to focus on the means by which atoms and molecules interact under ideal conditions. Engineering, on the other hand, is generally about making reactions happen at a high scale where other factors come into play, for example where the chemical conditions are either more severe than needed in the laboratory or are on a much larger scale than the conventional laboratory glassware. Under these circumstances, the desirable outcome of the science is often hampered by physical process parameters such as mixing, addition or removal of heat, and separation of products. These parameters become progressively more complex if the process involves more than one phase—a process that combines a gas with a liquid using a solid catalyst is much more interesting than a single-phase reaction such as gas–gas, liquid–liquid, or solid–solid reactants—unless there is a phase change part way through the process. The engineer would also ensure that the processes are operated safely under conditions that might involve high temperatures or pressures as well as corrosive or hazardous reagents.

The different technical terminology used by scientists and engineers or by scientists and engineers from different subdisciplines of science and engineering, can (and often does) result in miscommunication. Yet these professionals are required to work closely together in the process industry. There is a need to give scientists and engineers an introduction to the fundamental concepts and terminology of science and engineering and an understanding of why these concepts are necessary.

Under such circumstances there is obviously a need for reform in teaching science and engineering. As noted in the sections above, engineers need science and science needs engineering.

### 5.4.1 THE NEED FOR REFORM

The task of improving the quality of education (as well as closing or narrowing the knowledge gap) is linked to the task of reforming the higher-education sector (including the high-school sector) (Stensaker, 2013). With this goal in mind, several aims are necessary for improving the quality of higher education: (1) to improve the quality of education for scientists and engineers, (2) to hire only effective teachers/

professors for the science and engineering fields, (3) to initiate and maintain collaboration between preuniversity (high-school) teachers and university professors, (4) to disseminate specialized knowledge and skills in order to improve student development, (5) and to encourage students to explore novel areas in the context of filling the knowledge gap. In relation to Item 5, this should involve encouraging constructive thinking by the students while abandoning the current system of having the students parrot data back to find/guess/know the correct answer out of six possible answers to an examination question (Chapter 2). Educating students to parrot facts without understanding the background or derivation of the facts is to be deplored. Understanding the reasons for a certain equation is more educational than learning the equation by heart without understanding the whys and wherefores.

With reference to graduates in science and engineering, it is not adequate if the course and programs are redesigned merely to meet the demands of theoretical knowledge within a narrow area of specialization. Nor is it suitable to lower the standard required to complete and pass a course (Chapter 2). It sometimes happens that a teacher/professor decides that grading and marking methods must be changed to ensure that her/his favorites pass and the not-so-favored students fail or all of the class passes—the last option is used for self-gratification and to demonstrate the self-proclaimed excellent teaching methods of the teacher/professor. This is dishonesty in one form, and dishonesty in any form, let alone academic dishonesty, is a serious offense and violates all procedures by giving some students an unfair advantage (Speight and Foote, 2011).

There is a pressing need for universities to impart a sense of achieving quality work and to integrate the courses and programs to enhance the performance of prospective scientists and engineers.

### 5.4.2 ASSESSMENT AND ACCREDITATION

Administrators and faculty at science and engineering institutions (universities) need to be educated in terms of (1) the infrastructure, (2) resources, (3) faculty, and (4) programs of teaching and research. An immense gap in standards and facilities between many universities has led to serious concerns about the quality of education.

The students need to be able to access quality higher education, and this is a critical issue—putting in time and being awarded a degree on the basis of time spent is not quality education (Stensaker, 2013). In fact, issues of accessibility to quality education and improvement in the current quality of education are necessary aspects of the education of scientists and engineers and equal emphasis should be given to both. If this is not the case, institutions/facilities for the teaching of science and engineering will not be equipped to meet the challenges of the future. Quality improvement and the quest for excellence in educational teaching should be a continuous and perennial pursuit (Stensaker, 2013).

### 5.4.3 IMPROVING SCIENCE AND ENGINEERING EDUCATION

In all universities, teaching the scientific and engineering disciplines needs to be improved to achieve a twin goal: (1) the progress of higher education and (2) the development of science and engineering. Thus, various institutions should offer services to students from primary to higher educational levels—such centers may even be given functional autonomy within universities. Books and journals in libraries have immense value and the networking of resource libraries with the electronic transfer of information can provide the necessary browsing facilities to students and teachers.

The knowledge necessary for successful teaching in schools and universities lies in three domains: (1) deep knowledge of the subject matter (i.e., history and mathematics) and skills (i.e., reading and writing) that are to be taught; (2) expertise in instructional practices that cut across specific subject areas, or *general pedagogical knowledge;* and (3) expertise in instructional practices that address the problems of teaching and learning associated with specific subjects and bodies of knowledge, referred to as *pedagogical content knowledge* (i.e., a framework to understand and describe the kinds of knowledge that is necessary for a teacher to be in effective pedagogical practice in a scientific-enhanced or engineering-enhanced learning environment).

Beginner teachers/professors differ markedly from expert teachers/professors in their command of these domains and their ability to use them. For example, there will be differences in the types (and number) of examples and strategies used to explain difficult concepts to students. Differences in the range of strategies used by beginner teachers/professors will also be evident, as well as in the means with which they employ the strategies. Professional development courses that result in significant changes in practice will focus explicitly on the various domains of knowledge. Such courses will also encourage teachers/professors to analyze their own practice and provide opportunities for teachers/professors to observe experts and to be observed by experts as well as receive feedback from experts.

However, when experienced teachers/professors (with deeply ingrained practices) experience new models of teaching practice or when teachers/professors are asked to challenge what they think about the range of student knowledge and skill they can accommodate in a given classroom, the ingrained beliefs can detract from the acquisition of new knowledge. Thus, one aspect that can lead to improvement in the performance of teachers/professors is to encourage the teachers/professors to *unlearn* ingrained beliefs and practices that often work against the development of new and more effective practices.

Unfortunately, the disjunction between experience and expertise is an ever-occurring issue, but it is important to acknowledge that, in such cases, it will be necessary to recognize that experience does not always lead to expertise.

### 5.4.4   KNOW THE STUDENTS

In order to assist in the intellectual growth of students it is necessary to balance the support the teacher/professor can provide with the challenges posed by the course content. To many students (in high school and university) learning often seems a threatening task, which can cause the students to withdraw from learning. One of the tasks as a teacher/professor is to recognize the students' stage (or individual stages) of knowledge and to help bridge the knowledge gap, thereby successfully elevating the student(s) to the next level of knowledge.

While students may come to a new upper class with gaps in knowledge, there are also gaps in teacher/professor knowledge. The major gap in knowledge for any teacher or professor is when they do not know (or do not even bother to determine) the level of educational knowledge of the students. Assuming that the students are at a specific level of knowledge is one of the most fundamental errors made by any teacher/professor. Moreover, it is essential that a teacher has accreditation in the subject that s/he is teaching (Chapter 4). If the teacher/professor does not have any form of accreditation, serious action must be taken to either ensure that (1) the teacher is given the chance to get accreditation or (2) the teacher/professor is removed from that classroom.

To plan an effective course for students, it is essential to consider the following criteria in relationship to the prior education of the students: (1) prior knowledge, (2) intellectual development, (3) cultural background, and (4) experience and expectations. Teachers and professor may be filled with the best intentions but without knowing anything about student knowledge, the course may be presented in a vacuum of misunderstanding (Brookfield, 2006). The captain of the Titanic was filled with the best intentions, but without knowing anything of the patterns of flow of icebergs, these intentions did not come to fruition.

Students arrive at a class with a range of prior knowledge, skills, beliefs, and concepts that significantly influence what they notice about the environment and how they organize and interpret it. In turn, this affects their abilities to remember, reason, solve problems, and acquire new knowledge (Bransford et al., 2000). Furthermore, new knowledge is built on existing knowledge and the axiom " no knowledge then, no new knowledge now" is worth remembering. It is extremely important for a teacher/professor to determine what the students are likely to know coming into a course and how well the knowledge has been retained for future use. It is just as important for a teacher/professor to determine what the students are not likely to know and how well they need any lost (missing) old knowledge for future use. These issues can be resolved by a frank and honest conversation between the teacher/professor and a colleague who has taught the preceding course, (1) requesting a copy of the syllabus of the previous, course as well as (2) assignment titles, and/or (3) closed-book examination questions. The extent to which students have been required to actively achieve an acceptable status of knowledge—as evidenced by consideration of the above categories—can then be determined in terms of what the student has learned and how well they are able to apply the knowledge imparted to them.

Another aspect of the course-planning stage is for the teacher/professor to ascertain the major subject preferences and accomplishments of each student. Discovering this in advance can also help the teacher/professor determine how to effectively present new knowledge that is typically based on prior knowledge, as well as how to make the new material relevant and engaging. For example, using examples and/or illustrations and/or the use of objects in the correct contexts will certainly be an advantage for the students when it comes to connecting old knowledge to new material and to understanding the relevance of the new material to their own interests and future work.

New knowledge cannot be built effectively on a weak foundation, and thus it is important to determine where the prior knowledge of the student is *fragile* insofar as inaccuracies, naive assumptions, and/or misunderstandings exist within their knowledge base. The prior knowledge of students should be determined and can be achieved by the administration (in spite of many moans and groans from the students) of a simple diagnostic pre-test during the first week of class—adequate notice of the test (one or more days) can be given—which will identify areas of strong understanding or weak understanding (i.e., gaps in knowledge). This knowledge forewarns and arms the teacher/professor for the tasks ahead.

Throughout such exercises, it is worth remembering that students(of all ages) tend to see life in terms of right or wrong distinctions because (to the student) knowledge is unambiguous and clear, and learning a simple matter of information exchange. Students at this stage believe the role of the teacher/professor is to present facts, which they (the students) must remember and reproduce (regurgitate) when necessary—hence the use by some teachers/professors of the multiple-choice questions in test papers. If the teacher/professor presents facts that lead to ambiguous answers to problems, the students will be frustrated and confused. At about this point in their careers, students should be encouraged to realize that opinions differ, facts can contradict one another, and reasonable people can disagree.

Students should also be sufficiently encouraged to think for themselves and to question received wisdom. In addition, students should be able to understand that some perspectives have more validity than others, and that even the word of authorities should be analyzed critically (Baxter-Magolda, 1992). Furthermore, the students should begin to perceive the role of the teacher/professor not merely as a regurgitator of facts but also as a knowledgeable guide who is available to discuss the relative issues and mentor students through the knowledge-garnering process.

Finally, cultural differences among students (or even between students and the teacher/professor) can also affect course planning and presentation. The ways in which the roles of students and teachers are conceptualized may differ considerably from culture to culture. When students from different cultures share a classroom—or if the teacher/professor comes from a different culture than the students (as I discovered when teaching at a university in Iraq)—it is important to consider how cultural background can affect classroom dynamics and learning. It is necessary for the teacher/professor to (1) understand the types of challenges international students face, (2) explore issues that may affect students in the course, and (3) offer suggestions based on strategies that may be known by the teacher/professor that have been successfully employed. Such strategies that can help facilitate teaching and learning in a multicultural classroom serve the interests of all students—and the teacher/professor—regardless of cultural background.

### 5.4.5 Examinations and Standardized Achievement Tests

For decades, standardized achievement tests used in schools and universities have been the order of the day—tests take relatively little time to administer and the results are simple to report and understand. Often a single score, such as a percentile rank, standard score, or grade equivalent is reported for each student. In some cases, the marks are adjusted using the *bell curve* system (Chapter 2) which is used to describe the so-called normal distribution of the marks, after which the pass mark might be set according to the shape of the curve! And to make matters worse, standardized achievement tests have been promoted as objective measures of achievement—in spite of opinions to the contrary (Lipman, 1987; AASA, 1989; Mislevy, 1989; Wiggins, 1992; Barth and Mitchell, 1992)—meaning that the results are not affected by the personal values or biases of the person who marks the test and determines the scores.

For many individuals, an assessment system relying on objective measures of achievement appears entirely appropriate. Standardized achievement tests are promoted as scientifically developed tests that are valid and reliable measures of the scholastic performance of the students. Advocates of standardized testing have assumed that a student who had a specific knowledge (and the necessary facts) would also have an understanding of the course content. Consequently, each year the students at several grade levels in many school districts are tested using standardized achievement tests; invariably, the results show that most students perform far above the national average. It is worth remembering that the use of the word *average* is subject to misinterpretation but is endemic in the (school and university) education system. To the scientist or engineer, the word average means that if a person stands with one foot in a pail of ice water and the other foot in a pail of near-boiling water, they will be at an average temperature and all is well and there is no discomfort! And so it is with many standard achievement tests—there is considerable discomfort.

In fact, standardized achievement tests using a multiple-choice format are not effective in measuring complex problem-solving skills, divergent thinking, and collaborative efforts among students (Lipman, 1987). They also are ineffective in measuring communication skills and do not require full intense study of the subject matter through concentrated reading, since the response must be quick and nonreflective, and judgment, interpretation, and thoughtful inference (by the student) are not necessary to pass the test (Resnick and Resnick, 1989). In addition, a student who performs well on a standardized achievement test (with multiple choice answers) does not necessarily know more than his/her peers because the student can select (some student may be lucky in their selection of the answer) rather than construct an answer that is based upon true and well-learned knowledge. In fact, it is argued that most standardized achievement tests measure traditional basic skills and are not particularly effective in measuring the higher-order thinking skills that are crucial for the twenty-first century (Mislevy, 1989; Wiggins, 1992; Barth and Mitchell, 1992).

Indeed, serious reservations remain in relation to the continued heavy reliance on standardized achievement tests as the single or most important measure of how well student are doing in education (AASA, 1989).

## 5.5   A MULTIDISCIPLINARY APPROACH

In many universities, the prevalent view is that science has little contact with engineering. In government and industry the level of integration is greater, driven by the tangible benefits of close collaboration. Part of the (unjustifiable) reason is that scientists and engineers have a different but complementary knowledge base and, thus, they think and approach problems differently. This involves not just what is known but the means by which knowledge is applied. This diversity of thinking and perspective introduces some beneficial challenges into process development—relative to the laboratory scientist or process scientist, process engineers have a different perspective on scale-up, derived from their ability to predict using mathematical models and their understanding of equipment and manufacturability.

Process engineers (usually engineers who have been educated in the chemical engineering system) typically believe scale-up problems can be anticipated, providing there is sufficient understanding of the critical variable. For example, the number of theoretical plates affects batch distillations, and laboratory equipment usually has more than plant equipment. Therefore modeling the physical properties and equipment provides a better basis for understanding than several laboratory preparations. To support science-based scale-up it is essential to acquire data.

A process is dependent on science (usually, but not always, chemistry), which is independent of scale; and on physical transport phenomena —momentum, mass, and heat transfer—which are affected by scale. A linear reaction profile indicates a mass transfer limitation, and scale-up of such a reaction profile may not go as planned. This is increasingly important in multiphase systems or if the intrinsic (chemical) reaction rate is fast. Transport of reagents between phases or fast reactions may generate different concentration distributions at scale and affect yield and quality. A molecule does not know whether it is in a small flask (say, 250 ml) at the laboratory scale or a large reactor (100-U.S. gallon, 3,785,412 ml vessel) at the industrial scale—the molecule will simply behave as determined by the surrounding environment.

Unfortunately many scientists are not taught the intricacies and needs of scale-up from laboratory to commercial scale, and this is where the knowledge gap is not always bridged successfully in terms of the education of scientists and engineers. When moving from laboratory to plant it is not possible to keep all contributing parameters the same. In trying to keep one the same, others may be impacted detrimentally. The time required for (1) charging reagents to the reactor, (2) adjusting the temperature, (3) transferring materials, (4) removing gases and other volatile products, and (5) separating liquid phases or filtering solids could all change on scale-up due to equipment limitations or geometry. Changing vessel dimensions will change heat-transfer rates. As the surface area–to–volume ratios is reduced on scale-up, exothermic reactions run in semi-batch mode, under isothermal conditions that will require longer reagent addition times, which may affect quality.

In addition, mixing the reactants in stirred tanks is a complex scientific and engineering exercise and is affected by many equipment- and process-related parameters. It is therefore important to determine whether the process is sensitive to mixing before scale-up—laboratory vessels (the geometry of the vessel should mimic the geometry of the scale-up vessel) may need to be equipped with bafflers to encourage turbulent flow and better mixing with variable (to be determined) speeds of the agitator.

At the same time, consideration must be given to the reaction thermodynamics, which governs the nature and course of the reaction. For example, reaction equilibria can be explained by thermodynamics, which allows extrapolation of data into new compositions and prediction of the behavior of chemically similar systems.

Reaction thermodynamics can also be used to prevent azeotropic solvent exchange, which may cause the reaction to revert back to the starting reactants, with very few ending where they started; and can also allow a better understanding of a number of other operations dependent on phase equilibria.

From the above paragraphs, it is during an examination of the laboratory process for scale-up that there may be (will be) a fundamental difference between the process scientist and the process engineer. The scientist will always prefer to carry out more experiments (which is not always advantageous to a scale-up project) while the engineer will get to work with paper and sharpened pencil (or computer in these days of digitizing everything) using various theories and/or making various assumptions (which is also not always advantageous to a scale-up project).

*Vive la différence* is not always the way to proceed, and a heads-together session (rather than a head-butting session) session may be much more preferable. However, there is an optimal time for a scientist to collaborate with an engineer, but it is not definable. Nevertheless, it is far worse to miss an opportunity by holding off on a discussion than to talk things through early and get the most from the collective knowledge a scientist or engineer will bring to the table. This reflects the advantage of a not only a multidisciplinary approach but also a multidisciplinary education, where scientists understand some aspects of engineering and engineers understand some aspects of science.

In the twenty-first century, universities are faced with a critical problem—the link between teaching science and engineering and research in each discipline as well as the effectiveness of the methods applied. Many faculty members are not certain about what new developments may occur while the university is attempting to negotiate its way out of the present predicament. The model of the teaching-research university is not always viable, and the knowledge-frontier dilemma of the university pertains to the disconnect between teaching and research. Thus, the university is no longer capable of remaining the main institution for knowledge production.

The relationship between universities and other centers of knowledge such as research institutes has always been one immersed in controversies, where universities view research institutes as a burden upon the financial resources and the human resources that were really intended for the university. In part, the decline of the university was attributed to the rise of research institutes and (governmental and private) research laboratories. Another concern is that some universities have limited research facilities while the research institutes and research laboratories focus on research and there are minimal teaching duties, although mentoring is a necessary duty.

Together, universities and research institutes must revise their objectives and mandates. There is a need for communication and mobility between both parties. Indeed, the university system needs to acquire greater flexibility without compromising on its principles and practices in promoting knowledge.

## 5.6   STRATEGIES FOR CLOSING KNOWLEDGE GAPS

The existence of gaps in student knowledge is detrimental to student performance within institutions of education and then in the workplace. Schools and intuitions of higher education need a way to identify knowledge gaps, especially in critical areas such as (in the context of this book) science and engineering. It can seem like an overwhelming exercise, especially when this involves hundreds of thousands of students, located in different geographic regions, with different scholarly objectives. However, not only is it achievable, but it is also possible to reduce knowledge gaps, one student at a time.

The first step is to understand what the students do not know. Using preuniversity and university assessment techniques the initial gaps in knowledge can be uncovered. Ongoing frequent testing and self-assessment—carried out if possible in a stress-free manner—can be used to identify what learning has or has not been retained before it has an adverse impact on student performance. In fact, uncovering knowledge and application gaps is only the first part of the puzzle. Once it is known where knowledge gaps can be closed, it is almost assured that the learning is being absorbed and retained by the student(s).

The real measurement of both the effectiveness of education processes and the existence of knowledge gaps is achieved when a university evaluates knowledge retention. These evaluations can be used to identify what past learning students are able to apply to their current level of education and the measurable benefit to the students. Although difficult to measure, it is achievable through ongoing testing, student self-assessment, and professor/mentor assessment.

The one-size-fits-all programs that expose all students to the same broad stroke of information in the same manner should be discarded. The learning style and level of expertise of each student is unique, so are there are bound to be knowledge gaps. Personalization between the professor/mentor and the student is key to reducing these gaps.

Cognitive theory and development (a comprehensive theory about the nature and development of intelligent thinking) suggests that students are only able to retain about four items in short-term memory, making extended information-packed training sessions a potential minefield of knowledge gaps. By presenting knowledge in smaller, easy-to-assimilate pieces that are focused on what the student needs to know learning can be made easier to retain and apply.

Closing knowledge gaps means that students will be motivated to participate in the learning process on a continuous and attentive basis. Above all else, the key to finding and closing knowledge gaps is to ensure that learning and evaluation are done on a continuous basis, with the full attention of the student(s). The challenge in the classroom is that as soon as a learning event ends, information retention immediately begins to degrade, unless it can be continually reinforced until used. Interval reinforcement is a way to make it continuous.

The work of schools and universities is becoming more complex and demanding while the organization of schools and universities remains, for the most part, static and rigid. If pushed hard enough, a rigid structure will eventually break and hurt the students—this is the perilous state of American public education. The immediate cause of this situation is a simple, powerful idea dominating policy discourse about schools and universities—that students should be held to high, common standards for academic performance and that schools and universities and the people who work in them should be held accountable for ensuring that all are able to meet these standards.

Accountability schemes come in many forms, including high-stakes student testing and state takeovers of low-performing schools and universities. The word *accountability* also can refer to many things, including rules and procedures, or to the delivery of certain types of academic content (Glencorse, 2013). In the context of this book, the word is used only to refer to systems that hold students, schools, or universities responsible for academic performance, since this is the dominant form of accountability in education. Unfortunately, schools and universities were not designed to respond to the pressure for performance that standards and accountability bring, and their failure to translate this pressure into useful and fulfilling work for students is a danger to the future of public education.

The standards and accountability movement should be broad based politically and persistent over time (Glencorse, 2013). It must involve federal legislators, state legislators, advocacy groups, and professional organizations. The concept of such a movement stems from the basic belief that schools and universities, like other public and private organizations in American society, should be able to demonstrate what they contribute to the learning of students, and that they should engage in steady improvement of practice and performance over time. The movement expresses expectation that schools will face and solve the persistent problems of teaching and learning that lead to the academic failure of large numbers of students or failure (at the university level) to retain students and the mediocre performance of many more. Over time, if schools and universities improve, increased accountability will result in increased legitimacy for public education. Failure will lead to erosion of public support and a loss of legitimacy. However, with increased accountability, American schools and universities and teachers/professors are being asked to do something new: to engage in systematic, continuous improvement in the quality of the educational experience of students and to subject themselves to the discipline of measuring their success by the metric of students' academic performance (Glencorse, 2013).

The organization and culture of American schools and universities is, in most important respects, the same as it was in the late nineteenth and early twentieth centuries. Teachers and professors are still, for the most part, treated as solo practitioners operating in isolation from one another under conditions of work that severely limit their exposure to other teachers/professors doing the same work. The work day of teachers/professors is still designed around the expectation that teachers' work is composed exclusively of delivering content to students, not, among other things, to cultivating knowledge and skill about how to improve their work. Hence, the reasons for the existence of knowledge gaps in student learning.

The prevailing assumption is that teachers and professors learn most of what they need to know about how to teach before they enter the classroom—despite evidence to the contrary—and that most of what they learn after they begin teaching falls into the amorphous and difficult-to-define category of *experience*, which usually means lowering their expectations for what they can accomplish with students and learning to adjust to a school or university that is either noncommittal, unsupportive, or even hostile to their work. This limited view of what teachers/professors need to know and do demands little educational leadership from administrators. And, since administrative work currently has little to do with the content of teaching, much less its improvement, it may actually act to protect teachers and professors from various external intrusions on their isolated work.

The learning that is expected of teachers, professors, and administrators as a condition of their work also tends to be predicated on the model of solo practice. In order to advance in rank and salary, individual teachers, professors, and administrators are expected to accumulate academic credit for their respective levels of education (i.e., their position in the same system that they are currently propagating), any or all of which may be totally unconnected to their daily work. Most workplace learning also mirrors the norms of the institution in which they work and takes the form of information about policies and practices delivered in settings disconnected from where the work of the institution is actually performed.

Perhaps the irony of the present American education system (and the systems of many countries for that matter) is that schools and universities are hostile and inhospitable places for learning. They have been this way for some time. In the current situation, the advent of performance-based accountability may ultimately undermine the legitimacy of public education, and actions must be taken to change the way schools and universities perform their respective duties to the community of students (Glencorse, 2013).

## 5.7 THE FUTURE

While some professors may not have changed their course notes for 20 years (Chapter 3) in real terms the actual content of scientific and engineering courses has changed considerably. Until World War II, the science and engineering curricula at most universities was generally sensible, practical, and emphasized experimental design and engineering design, respectively, with practical skills, and the courses were taught by faculty with experience and ongoing activity in science and engineering practice (Duderstadt, 2008). This is not the case in the modern university, where the focus has shifted to more esoteric subdisciplines with more theory than practice and a greater emphasis on basic (often theoretical) research. As a result, most undergraduate science and engineering programs in the modern university offer only lip service to experimental design, technical writing, and professional ethics so as to pass the muster of the accreditation board (such as ABET, the Accreditation Board for Engineering and Technology).

Clearly the science and engineering curricula need a major overhaul. To some degree, this will require modernizing the approaches to science and engineering instruction, for example, recognizing that discrete rather than continuous mathematics is the foundation of the digital age, that biology is rapidly becoming as important as physics and chemistry, and that new scientific concepts and tools have made obsolete much of the traditional curriculum (Duderstadt, 2008). Beyond these technical changes, new science and engineering curricula must reflect a broad range of concerns—environmental, political, social, and international, as well as the ethical ramifications of decisions. Although scientific and engineering courses would continue to be the core of their respective education programs, the economic, political, social, and environmental context of scientific and engineering practice must be addressed.

In a scientific and engineering economy that is based on knowledge, the transformation of knowledge into products, processes, and services is critical to competitiveness, long-term productivity growth, and the generation of wealth. Preeminence in technological innovation requires leadership in all aspects of science and engineering to convert discoveries to practical applications. This cannot be accomplished if there is a knowledge gap.

The scientific and engineering education received in universities (with suitable preuniversity preparation in schools) must give scientists and engineers the necessary skills to create and exploit knowledge and technological innovation. Scientific and engineering professionals must have the ability to translate knowledge into innovative, competitive products and services. In fact, to compete with talented scientists and engineers in other nations with far greater numbers and with far lower wage structures, American engineers must be able to add significantly more value than their non-American counterparts abroad through (1) a greater intellectual span, (2) the capacity to innovate, (3) entrepreneurial zeal, and (4) the ability to address the necessary challenges of a project.

From the items presented in the preceding paragraph, the key to producing such top-class scientists and engineers is for American schools and universities to provide the opportunity to significantly broaden the educational experience of science students and engineering students, with the professional job marketplace showing a willingness to help meet these objectives. Essentially, the scientific and engineering professions must achieve the status and influence of other learned professions such as law and medicine. Engineering practice requires an ever-expanding knowledge base requiring new paradigms for scientific and engineering research that better link innovation and development with the real world. There should be no tolerance or office space for the faculty member who uses the university to hide away from the world and whose only activity is taking a 2-h coffee break with his/her colleagues.

In order to establish scientific and engineering practice as a true learned profession, similar in rigor, intellectual breadth, preparation, stature, and influence to law and medicine, there must be no knowledge gaps, or at least such gaps as there are must be able to be quickly and effectively plugged. To accomplish this goal, the nature of basic and applied teaching must be changed. New paradigms that better address compelling priorities will need to evolve—at the soonest possible time. This will allow the evolution and adoption of a systemic, research-based approach to innovation and the continuous improvement of science and engineering education, recognizing the importance of diverse approaches—characterized by quality—to serve the highly diverse scientific and engineering needs of the present and the future.

There is little doubt that the current sequential approach to education in science and engineering education, in which the early years are dominated by various nonscience and nonengineering courses, with the scientific and engineering content deferred to the upper-class years, discourages many capable students. Compounding this is the fragmentation of the current curriculum, consisting of highly specialized and generally unconnected and uncoordinated courses, where relationship of one course to another and to education in scientific and engineering disciplines is rarely explained. Students have little opportunity to find out what engineering is all about until late in their undergraduate studies. There is little effort to relate the curriculum to career and professional development opportunities during the early years of an engineering education (Duderstadt, 2008). It is not unusual to find students wandering into counseling and placement offices in their senior year, still trying to find out what they are majoring in and what they can do with a degree in science or engineering.

Competence in science or engineering can only be accomplished if scientific and engineering professional societies work with science and engineering leadership groups. Schools and universities should strive to create a guildlike culture in the science and engineering professions, similar to those characterizing other learned professions such as medicine and law, which aim to shape rather than simply react to market pressure.

Companies of the twenty-first century require scientists and engineers develop new skills and competence that address the challenges and opportunities of twenty-first-century business. This has particularly serious implications for the future of science and engineering, since not only must engineers develop the capacity to work with multinational teams and be internationally mobile, but they also must appreciate the great diversity of cultures characterizing both the colleagues they work with and the markets they must compete in. Furthermore, the scientist and engineer face the additional challenge of competing internationally with scientists and engineers of comparable skills.

The modern international (or multinational) corporation conducts strategy, management, and operations with and within other countries. The multinational organization has evolved far beyond a collection of country-based subsidiaries to become instead an integrated company of international components, which also conduct research, development, manufacturing, and sales. Furthermore, national borders are of declining relevance to the business practices of the international corporation and, in some cases, such corporations may even show a diminished level of loyalty to the country of origin as more attention is paid to business activities in the countries where new markets and business opportunities exist (Palmisano, 2006).

With this in mind, universities should establish graduate professional schools of engineering that would offer *practice-based* degrees at the postbaccalaureate level as the entry degree into the scientific and engineering profession (Singh, 2011). In addition, undergraduate science and engineering should be reconfigured not only as academic disciplines but also disciplines that provide more flexibility for students so that they (the students) can benefit from the broader educational opportunities offered by a more comprehensive university, with the goal of preparing the students for a career with continued learning (to bridge the knowledge gap). In fact, it is essential that the scientific and engineering professions adopt an approach to lifelong learning for practicing scientists and engineers similar to those in medicine and law—a course here and there at an annual or semiannual meeting may be of insufficient educational interest for the maturing (and mature) scientist and engineer. Such a change in the educational paradigm will require not only a significant commitment by educators, employers, and professional societies but possibly also (whenever necessary) additional licensing requirements in some fields of science and engineering.

Finally, a word about teaching faculty—the role of these persons must not be ignored or forgotten in any commentary about the future of science and engineering courses. The faculty members of the various university science and engineering departments are often (or typically) quite different from the faculty members of most professionally oriented schools and colleges, since they generally have lesser experience or fewer ongoing activities in professional practice. This can be borne out by discreet enquiries made of the local university as to how many professors have ever left their office for any professional activity other than teaching-learning activities, or the number of professors who regularly attend a social function where one can stand around holding a bottle (or a glass) of beer and strive to look important and intelligent at the same time. The answers to such enquiries may be very surprising to many observers.

In other cases, the strong research focus of many scientific and engineering university departments has led to a cadre of strong scientists and engineers who are quite capable of generating new knowledge but who are relatively inexperienced in applying this knowledge in professional practice. Furthermore, scientific and engineering faculty members are judged and rewarded by criteria (such as publication in various journals and grantsmanship) that often bear little relationship to the needs of the real world or to professional practice. In fact, in the assessment of academic scientists and engineers in the category of faculty evaluation (Section 5.3, this chapter) is not only absent in promotion and reward criteria, but is also frequently ignored or even discouraged. After all, the university's dean of science and dean of engineering as well as the vice president and president arose to their present appointments through a similar system. The faculty reward system recognizes teaching, research, and service to the profession, but it gives little recognition for developing a marketable product or process or designing an enduring piece of the nation's infrastructure.

Such detachment from professional practice and experience (i.e., the real world outside of the hallowed halls and walls of academia) is the norm in the lives of university scientific and engineering faculty.

## REFERENCES

AASA (American Association of School Administrators). 1989. *Testing: Where We Stand.* American Association of School Administrators, Arlington, VA.

Avenier, M.J. and Bartunek, J. 2010. Bridging a supposedly unbridgeable gap: Elaborating scientific knowledge from and for practice. Cahier de Recherche No. 2010-02 E4. Unité Mixte de Recherche CNRS/Université Pierre Mendès-France, Grenoble, France.

Barth, P. and Mitchell, R. 1992. *Smart Start: Elementary Education for the 21st Century.* North American Press, Golden, CO.

Baxter-Magolda, M.B. 1992. *Knowing and Reasoning in College: Gender-Related Patterns in Students' Intellectual Development.* Jossey Bass, Wiley, San Francisco, CA.

Bransford, J.D., Brown, A.L., and Cocking, R. (eds) 2000. *How People Learn: Brain, Mind, Experience, and School.* National Academy Press, Washington, DC.

Brookfield, S.D. 2006. *The Skillful Teacher: On Technique, Trust, and Responsiveness in the Classroom*, 2nd edn. Jossey Bass, Wiley, San Francisco, CA.

Donohue, G.A., Tichenor, P.J., and Olien, C.N. 1975. Mass media and the knowledge gap: A hypothesis revisited. *Communication Research*, 2(1): 3–23.

Duderstadt, J.J. 2008. *Engineering for a Changing World: A Roadmap to the Future of Engineering Practice, Research, and Education*. The Millennium Project, University of Michigan, Ann Arbor, MI.

Gaziano, C. 1997. Forecast 2000: Widening knowledge gaps. *Journalism and Mass Communication Quarterly*, 74(2): 237–264.

Glencorse, B. 2013. Testing new tools for accountability in higher education. In: *Global Corruption Report: Education* (Chapter 4.14). Routledge, Taylor & Francis Group, Abingdon, Oxford.

Houlton, S. 2013. *Learn to Speak Engineering*, p. 67. Chemistry World, The Royal Society of Chemistry, London.

Hwang, Y. and Jeong, S-H. 2009. Revisiting the knowledge gap hypothesis: A meta-analysis of thirty-five years of research. *Journalism and Mass Communication Quarterly*, 86(3): 513–532.

Lipman, M. 1987. Some thoughts on the formation of reflective education. In: J.B. Baron and R.J. Sternberg (eds), *Teaching-Thinking Skills: Theory and Practice*, pp. 151–161. W.H. Freeman, New York.

Mislevy, R.J. 1989. *Foundations of a New Test Theory*. Educational Testing Service, Princeton, NJ.

Neuman, S.B. 2006. The knowledge gap: Implications for early education. In: D.K. Dickinson and S.B. Neuman (eds), *Handbook of Early Literacy Research* (Chapter 2.29-40), vol. 2. Guilford Press, New York.

Neuman, S.B. and Celano, D. 2006. The knowledge gap: Implications of leveling the playing field for low-income and middle-income children. *Reading Research Quarterly*, 41(2): 176–201.

Palmisano, S.J. 2006. The globally integrated enterprise. *Foreign Affairs*, 85(3): 126–136.

Resnick, L.B. and Resnick, D.P. 1989. *Assessing the Thinking Curriculum: New Tools for Educational Reform*. Learning Research and Development Center, University of Pittsburgh and Carnegie Mellon University, Pittsburgh, PA.

Salmi, J. and Helms, R.M. 2013. Governance instruments to combat corruption in high education. In: *Global Corruption Report: Education* (Chapter 3.2). Routledge, Taylor & Francis Group, Abingdon, Oxford.

Singh, V. 2011. Cooperative education: Making the transition from knowledge society to knowledge economy. In: *Proceedings of the Association of Caribbean Higher Education Administrators (ACHEA) Conference*. July 7–9, University of the West Indies, Cave Hill Campus, Barbados.

Speight, J.G. and Foote, R. 2011. *Ethics in Science and Engineering*. Scrivener Publishing, Salem, MA.

Stensaker, B. 2013. Ensuring quality in quality assurance. In: *Global Corruption Report: Education* (Chapter 3.5). Routledge, Taylor & Francis Group, Abingdon, Oxford.

Tichenor, P.A., Donohue, G.A., and Olien, C.N. 1970. Mass media flow and differential growth in knowledge. *Public Opinion Quarterly*, 34(2): 159–170.

Wiggins, G. 1992. Creating tests worth taking. *Educational Leadership*, 49: 26–35.

# 6 The Complete Scientist and Engineer

## 6.1 INTRODUCTION

A university is a community of students and teachers committed to the pursuit of learning, the accumulation of new knowledge, the transmission of new knowledge to succeeding generations, and the development of new knowledge. A university combines teaching, research, and discovery, as well as (in some cases) community service. These principles give rise to a collection of scientists and engineers who can give a university a unique outlook. Hence, good students of the science and engineering disciplines must commit to being lifelong learners.

Over a century ago, the German universities first arrived at a consensus that teaching and research are complementary activities, and that the maximum success in each discipline area is only attained within any environment (especially in an academic environment) in which both subject areas are fostered and encouraged. In order to accomplish advances in science and engineering, it is essential that faculty and students within the colleges of science and engineering work together with their peers and with each other. Furthermore, there must be a close link between undergraduate and postgraduate work and between students and academic staff who have a wealth of experience in the respective fields at different levels of the university. Cooperation and collaboration for the pursuit of knowledge is a means to strengthen the quality of the education of scientists and engineers, which may also impart valuable lessons for the workplace as well as contribute to the development and transmittal of new knowledge to posterity and thus meet the needs of succeeding generations.

Graduate scientists and engineers have traditionally been educated and prepared for employment positions in which the ability to perform original research is the skill of highest value. The traditional positions include research-intensive occupations in (1) academia, (2) industry, and (3) government laboratories, but scholarship and research—especially basic research in which new concepts are tested and developed—may constitute the primary focus of employment in academia while applied scientific and engineering research is more pertinent in the domains of government laboratories than in industrial laboratories.

However, different disciplines and subdisciplines of science and engineering vary widely with respect to employment patterns in research, in terms of the development of motivation, personal skills, and written communication. Issues that follow from this include the translation of theoretical knowledge into practical experience as well as the development and protection of intellectual property and its role in education. It is the purpose of this chapter to present each item as it pertains to the education of scientists and engineers.

## 6.2  MOTIVATION

Motivation is the act or process of a person having a reason for taking an action and is aligned (in the current context) with the condition of being ready and/or eager to act or work. This definition has been operative and obvious for many decades, well into historical times, especially in the innovations and development leading to advances in science and engineering (Singer et al., 1954–1959; Forbes, 1958a,b, 1959, 1964; Griffin, 2011).

Because of the layoffs that have occurred in the past four decades as well as a variety of other reasons, people are rarely loyal to companies or organizations but are loyal to one another, especially colleagues they can trust. Motivation can be seen in the form of rewards for innovations that the organization can use (in the academic world) to attract research funding or (in the commercial world) to receive an innovator's bonus. In science and engineering, there are rarely any big financial rewards unless the scientist or engineer initiates and develops a unique innovation in the form of intellectual property, which belongs to the scientist or engineer alone and not to the company or organization where the scientist/engineer is employed. Even then, legal issues related to the conditions under which the invention was first perceived as a concept and then proven may become major issues and proceed through the courts over a period of several years.

Assuming that a scientist or engineer does not tread on to the pathway to riches through an invention that is his or hers alone, competition for jobs and research funding is fierce and can make this a cut-throat world or dog-eat-dog world with more and more postgraduates (typically postdoctoral graduates) seeking limited positions and funding. Thus, a graduate who wishes to be considered for a position in scientific or engineering research has to be top level, well-educated, and dedicated. Because the bar is set so high, a high failure rate exists when, say, doctoral graduates compete for research positions. Furthermore, those who fail to be offered such appointments are destined for (1) further postdoctoral work or (2) an unwanted position at a lower level warranted by his/her capabilities. On the other hand, successful candidates who receive a prestigious research appointment (insofar as s/he applied for and was awarded the position) often labor for decades without much in the way of applause or recognition. Scientists and engineers also rarely enjoy the lavish perks of corporate life, such as (1) business-class or first-class travel, (2) four-star and five-star hotels, (3) dinner with clients—in many cases the scientist or engineer is relegated to meeting the client in a meeting or just prior to making a presentation to the client—and (4) the unwritten social perks of corporate life. These perks are omitted for the entire career life of the scientist or engineer, which can be a rude awakening for many scientists and engineers but, with careful mentoring during the university years, the experience of this awakening can be less rude and more understandable (NAS, 1997).

During the early years of graduating with a doctorate, one of the issues facing scientists and engineers is the general lack of motivation in (1) the academic system or in (2) the corporate system. While both systems can be a joy or a tragedy for the young scientist or engineer, there are many travails that are worthy of comment here. Not all academic educational institutions and not all corporate research institutions follow the trends presented in the next paragraphs, but many (in spite of anticipated vociferous denials from some readers) do.

On the one hand, in the academic system, the doctoral graduate scientist or engineer (or the postdoctoral scientist or engineer) considers himself/herself fortunate to receive an appointment at the assistant professor level. S/he is led to believe that the appointment is the best thing since sliced bread and that s/he (the appointee) is lucky to have been selected from the 100 or so applicants—in reality there may have been less than 20 (often less than 10) other applicants.

On the other hand, in the corporate system, the doctoral graduate scientist or engineer (or the postdoctoral scientist or engineer) considers himself/herself fortunate to receive an appointment at the junior scientist or junior engineer level. S/he is also led to believe that the appointment is the next best thing since sliced bread, and that s/he (the appointee) is lucky to have been selected from the 100 or so of other applicants—in reality (as in academia) there may have been less than 20 (often less than <10) other applicants. In some cases, the attention of the young scientist or engineer may have been solicited by corporate recruiters and s/he may be one of a kind, but a "one of a kind" that is necessary for the corporation. In fact, in the environs of the corporate halls of power and once ensconced within the corporate structure, the new recruit may also be considered to be a very general dogsbody (available to work on any assignments given to him/her) but with benefits—the benefits being that the young scientist or the young engineer is a source of (unacknowledged) stimulation and an idea factory for the group boss.

Corporate life provides another aspect of training for the young scientist and engineer—s/he will be appointed (without being asked or consulted for preferences) to a group studying a subject area that may or may not fit within the capabilities of the young recruit. Thus, the new recruit will find himself/herself invited to group meetings and to committee meetings. Feeling that s/he is to receive recognition for her/his educational talents, the recruit will feel wanted and important. Such meetings are invariably chaired by a senior faculty member or a senior corporate scientist or engineer. At some stage of the meeting, ideas relevant to certain issues may be requested by the chairperson. Being an eager beaver and wishing to make himself/herself known, the new recruit may place one or more ideas on the table.

Worthwhile or not, such ideas are typically received in one of two ways: (1) They are rejected outright, and the recruit is made to feel redundant or that s/he has just arisen from Dr. Frankenstein's table and is completely out of touch with corporate reality, or (2) the Chairperson feigns agreement with the recruit's idea and suggests to the group that they give that idea consideration, without saying yes or no. This latter approach also means that the idea can be shot down in flames by cronies of the chairperson in the privacy of the chairperson's office. On the other hand, if the idea does not spiral to earth and end in a fiery crash, and is instead successful, it is only after the fact that the recruit may learn that his/her successful idea has been claimed by the chairperson as his/her property—failure has one owner, success has many.

In either case, the new academic or corporate recruit is then subject to the whims and fancies of academic life or corporate life, including the whims and fancies of an official or unofficial mentor who may not give much thought to the needs of the new recruit. Indeed, in academia, the new recruit becomes the dogsbody for every senior professor who needs a task completed—a task that may fall within the capabilities of the professor but which the professor considers beneath his/her dignity (and exalted position in life) to carry out. Immediately, the new recruit learns that s/he must bow to the will of others or be refused tenure when the judgment day arrives. Some will not accept such a punishing and demeaning learning curve and depart for greener pastures—if the pastures are available and if there is sunlight on the pastures to sustain the greener life. If not, the recruit will work to fit into the pattern and eventually, after learning to say yes to the right persons (i.e., the pseudo–powers that be), will receive promotion to a tenured position—in some cases this would be perceived as corruption, but not in other cases (Shola Omotola, 2013). In the corporate system, the new recruit who is seen to be ambitious (who has never been such a recruit?) and who has provided several new ideas for his immediate supervisor(s) must also learn to say yes to the right persons—and every morning s/he may have to practice 100 different ways of saying yes in front of the mirror—before due recognition is given.

The real issue then becomes the means by which young scientists and engineers can be motivated, both in the academic system and in the corporate system. Scientists and engineers receive little (if any) form of leadership training and are not taught how to manage teams—they learn the hard way, by experience, and must pay attention to personal performance, which not only includes successes but also failures, providing that the failures have not driven the scientist or engineer into a hole so deep that s/he cannot get out of it.

Every scientist or engineer who does not eye the management chain as a means of climbing the academic ladder or the corporate ladder wants to be successful in his/her profession. Some tasks may be too daunting and help may be required—the key is to finding a colleague who can be trusted and is not likely to abscond with the idea of task resolution and claim it as his/her own idea. On the other hand, the scientist or engineer has to decide if the daunting task can be broken into several generally less daunting and achievable subtasks. If so, trust of a colleague may not be an issue and the young scientist or engineer, based on his/her experience to date in either the academic world or in the corporate world (see above), has provided a form of self-motivation. In fact, achieving each subtask as a result of one's own efforts builds confidence which, in turn, drives motivation, especially self-motivation. At this stage, the immediate supervisor/mentor/boss is then provided with a *fait accompli* in the form of a written memorandum and/or project report that can offer idea protection, leaving the young professional with a feeling of inner confidence.

If the supervisor/mentor/boss has behaved in a responsible manner s/he will have advised the young scientist/engineer to think about the task without rushing to solve problems—unless failure to solve the problem will result in plant shutdown or severe loss of revenues, which is unlikely in academia—and the supervisor/mentor/ boss will offer any other wise counsel to the recruit. Both confidence and expertise have been the results of such an exercise, and it has initiated good relationships. Furthermore, the skills learned by the young scientist or engineer will be valuable and s/he will be motivated to work with the supervisor/mentor/boss and develop further successful projects.

At group meetings dealing with specific or nonspecific science or engineering topics, the supervisor/mentor/boss should, at the beginning of such a meeting ask who is absent, which shows s/he notices and cares (or, at least, pretends to care). By celebrating birthdays, discussing movies or theater or the news, s/he gives the meeting attendees time to get to know each other as people, and not just scientists or engineers. Then, when a project presentation is on the meeting agenda, each of the other attendees should adopt a constructive role (nonconstructive or destructive roles should be forbidden but rarely are): reviewer, referee, and provider of the funds for the work, who all wish to ensure themselves that the presentation and writing is of high quality (Walters and Walters, 2011; Laplante, 2012). The goal is to create a sense of connectedness so the attendees wish to help each other. The group starts to think like a community, and the culture of helping one another not only provides the group members with the correct context and empathy for particularly difficult projects but also acknowledges successes.

The most difficult part of motivating scientists and engineers is to offer help when needed and to decide which part of a project is the most difficult and should be the object of focus. A project that is too hard cannot be solved until or unless new thoughts and ideas leading to new research provide the information needed to crack it. A problem that is too easy does not yield valuable insights—but it can be asked how often a supervisor/mentor/boss chooses such a project for support because the support will make him/her (the supervisor) look better (read: *successful*) in the eyes of upper management? Often!

A motivational supervisor/mentor/boss will ask scientific and engineering members of the team for input on possible relevant projects at a group meeting. S/he will then identify the talents and passions of the group members and consider where those overlap with the objectives of the group. In other words, each assignment is a rich intersection of personal capabilities with professional needs. There's no point asking (making) faculty or corporate staff members do work they do not care about—to do so is to reduce the value of the hiring process, and the energy of passionate scientists and engineers cannot be harnessed. The greatest motivational aspect of work for scientists and engineers is to foster personal and professional development. While this is a great part of the motivational process, promotion (genuine promotion for success) and money also help.

While the above paragraphs have focused on motivating scientists and engineers in the academia and in business corporations, it is now time to focus on scientists and the engineers in the realm of government laboratories. Scientists and engineers in governmental laboratories and governmentally funded research institutions are typically motivated by the ability to do high-quality, curiosity-driven research and demotivated by lack of feedback from management, difficulty in collaborating with colleagues, and constant review and change (NAP, 1995; Jindal-Snape and Snape, 2006; Hazelkorn, 2011). Tangible motivational rewards such as salaries, incentive schemes, and prospects for promotion are not always considered to be motivating factors by many scientists and engineers, especially when the rewards are doled out without any apparent merit, rhyme, or reason.

In fact, many scientists and engineers consider that after one or two promotions up the technical ladder, further promotions might be few and far between, and that after the ascending first rung or two of the ladder, they may never be promoted to higher elevations in the scientific or engineering ranks. In cases such as these, supervisor/mentors/bosses should focus their motivational efforts on addressing the working environment, especially the removal of negative factors and hindrances to motivation, instead of attempting to introduce incentives that are of little value and which, in fact, detract from moving ahead.

Motivational schemes based on financial rewards may have some, but have generally little, impact on motivation and alternative methods of motivating scientists and engineers should be considered. Motivational rewards that may be more highly valued could include (1) time and resources to pursue own research interests, (2) funds to attend international conferences, (3) and investment in physical resources such as laboratory refurbishment and new equipment.

However, on a cautionary note, such rewards can have a reverse effect. When scientists and engineers feel that material rewards are being dangled before them like carrots on a stick before a donkey, they can feel that they are being externally controlled, and this will have a dampening effect on motivation. In addition, as happened in one government organization, one of the scientists was singled out (correctly or incorrectly is not the issue) by management to attend almost one national/ international conference per month to make presentations on the status of the work. While this may have been motivational for the scientist, it was extremely demotivational for other members of the scientific and engineering communities of that organization. When this preferential motivation was stopped by new management, the scientist was demotivated but the remaining scientist and engineers who then received travel allowances to attend national and international conferences were energetically motivated, and their work output and work quality improved. In fact, creativity flourishes when scientists and engineers know that rewards and recognition will follow from good, creative efforts—without being told constantly about exactly which rewards will follow from which actions. Work-related rewards, such as opportunities to pursue other, potentially exciting work, are especially effective as motivators.

To extend this idea to scientific and engineering research, it can be noted that while there are a few awards around, science and engineering organizations do offer some noteworthy extrinsic rewards. But the problem in science and engineering is that the *right* kind of reward is only too rare. Typically, scientists and engineers in training need confidence-building experiences, and a record of consistent attainment will be rewarded with career advancement—with new opportunities, additional resources, technician assistance, and an increasing independence of thought leading to an increase in innovation. However, depending upon the research and funding climate, only a fraction of scientists and engineers receive the rewards they deserve.

In summary, scientific and engineering research teams can, in addition to professional scientists or engineers, also comprise postdoctoral associates, graduate students, and research assistants—each of these latter individuals have different career goals and are highly intelligent and prefer to have research autonomy. Managing such a diverse culture in a research team teams is like herding North American bison—if you (the reader) have not tried this herding bison exercise, please do not, because a bison, in the best mood the animal ever had on the best day of its life, can only be classed as surly and is extremely unpredictable.

## 6.3  INTERPERSONAL SKILLS

Interpersonal skills are a necessity in scientific and engineering research. The diverse culture of the teams in any academic institution, corporate research organization, or governmental research organization precludes many beneficial interpersonal interactions. In fact, many of the scientific and engineering team members may have the same mental interpersonal aptitudes as the unfriendly and unpredictable bison of the preceding paragraph.

Thus, in order to develop interpersonal skills, the scientist or engineer must (unlike the bison) be able to communicate, lead, network, and show teamwork with other professionals, including his/her peers, managers, and technical staff.

### 6.3.1  COMMUNICATION

Good communication is the essence of the life of any scientist or engineer, whatever his/her role in the workplace. The basis of good communication is very simple: it is speaking or writing clearly such that any message the scientist or engineer (the *sender*) transmits to someone else (the *receiver*) communicates precisely the reason for and meaning of the message. As well as the detailed content of the message, there must be thought given to the language used so that the receiving person will not misunderstand any words or phrases in the message. The sender must also be aware of the manner (the *medium*) by which the message is delivered—such as conveyance by word of mouth (i.e., through a face-to-face meeting, telephone, or e-mail. Hiding the reason and meaning of the message in a format that requires the services of the Bletchley Park code breakers defeats the object of the exercise.

Briefly and by way of explanation, Bletchley Park (45 min by rail northwest of London, United Kingdom) housed an organization called the Government Code and Cypher School (GC&CS), where the members of the group studied and devised methods to enable the Allied forces to decipher the military codes and ciphers that secured communications by and among Axis nations.

Even when the medium of message transmittal is carefully considered and the message delivered, achieving successful communication is often rather more difficult than it seems. Any single message, especially one of many sent and received during a busy working day, can be subject to distractions such as external noises or the thoughts of the receiving person dwelling on other matters. As a result, the sender may not be concentrating fully on the message s/he is intending to send and the receiver may not be concentrating on the message that is being received.

In terms of the communication skills required for leadership, it is important for the leader to give just as much of his/her attention to a message being received as to one being sent. Listening, for example, as well as being a method for gathering information, can convey the interest of the team leader (or mentor) in what the other person is saying. This in itself can have a positive effect on a relationship with others and, if the scientist or engineer is a manager, the motivation of a team. Asking questions for clarification of the detail of the message can also convey that the team leader has understood the message and that s/he wishes to respond to it appropriately.

This is certainly not the time for sarcastic, or "I-know-it-all" or "I-told you-so" responses, as is often the case—such responses only serve to discourage team members or students and it is a lesson for the young scientist or engineer in how not to respond.

### 6.3.2 LEADERSHIP

Leadership is the process of successfully influencing the activities of a group toward the achievement of a common goal. A leader has the ability to influence others through qualities such as personal charisma, expertise, command of language, and the creation of mutual respect. As well as requiring strong communication skills and personal skills, leadership uses the background skills of mentoring, decision making, delegation, and motivating others.

Although the young scientist or engineer is not always called upon to act in a leadership role, such a call can happen and the enthusiastic scientist or engineer must be prepared. There may (should) have been the need for the scientist or engineer to show leadership in working with a technician, remembering that the laboratory skills of many seasoned (mature) technicians are far superior to the budding laboratory skills of the young scientist or engineer. In academia, this may be different, but there are many undergraduate or (especially) graduate students with laboratory skills that can far exceed those of the professor/mentor.

There is an interesting, often puzzling, series of cultural barriers in science and engineering that tend to limit how scientists and engineers value leadership and the actions that are considered to be the actions of a leader. Some are obvious leaders—(1) persons who can capture the attention of team members by showing consideration or (2) persons who destroy a team by driving the project into the doldrums and failure but still deny to upper management that it was their fault and reap the benefits of success while others get the blame. Then there are those in academia who tend to lead indirectly, by example, but not all examples are good shining examples of exemplary behavior. Typically, scientists and engineers seem to associate leadership authority with those persons who have made great strides in science and engineering. Scientists and engineers tend to undervalue (even ignore) other types of leadership and prefer to be led by someone with technical credentials that are often referred to in superlative terms—true or not!

On the other hand, for some reason, scientists and engineers assume (incorrectly) that leadership is something that is hidden within the coiled helixes of the DNA molecule and needs to be extracted or is not available for extraction and use—that is, the prevailing thought is that the scientist or engineer is either born with leadership qualities or is born without such qualities. If any scientist or engineer wishes to witness the power of leadership training, s/he should consider the men and women in the Armed Forces. The recruits (young men and young women in the 18–20 age group) often come from extremely humble and sometimes difficult backgrounds, yet the focus on leadership and training in the military turns them into remarkable leaders. The military programs on leadership are second to none.

On the other hand, nonmilitary leadership programs do not often provide the teaching necessary for competent leaders in science and engineering. The objective of any leadership program in academia, government, or industry must be to develop the next generation of scientists and engineers whose leadership contributions will serve to strengthen and unify the science and engineering communities at national and local levels. The program must prepare scientists and engineers to assume leadership roles in the global science and engineering organizations and communities by offering advanced strategic training that helps develop any latent leadership skills. If the program cannot do this, the academic institution or organization sponsoring the program is merely presenting lip service without foundation or substance to the attendees.

### 6.3.3 Networking

Networking is the ability to actively seek, identify, and create effective contacts with others, and thence to maintain those contacts for mutual benefit. In addition to strong communication skills and personal skills, networking uses the background skills of network building and of motivating others.

Finding new research partners can be a challenge for scientists and engineers, as it often requires them to step outside of their daily commitments. But it is important, since meeting scientists and engineers from other disciplines can spark a new research idea or open the door to a solution to a problem that has seemed intractable. And yet such personal interrelationship activities are not always possible. Graduate students in academia are not always encouraged to network. The concept of networking insofar as projects are discussed with others may be forbidden by the professor/mentor who has the necessary paranoia or fear (to some extent, justifiable in these days of competitive funding for projects) that his/her project may be jeopardized if unpublished information is divulged to others in the same field. This paranoia or fear may even apply to other workers in the same university.

On the other hand, in industry, such personal interrelationship activities are always possible, although not with outsiders. Scientific researchers and engineering researchers in the commercial world are often encouraged to network with other scientists and engineers on an intracompany interdivisional basis. The concept of this form of networking may have to have the blessing of the respective managers and/or vice presidents but is usually welcomed. The work is for the good of the company (as opposed to being for the good of the individual professor) and more often than not written and witnessed patent memoranda to establish the inventor of the concept are in place. Thus, the accompanying paranoia or fear is not typically present in the environs of intracompany networking. This paranoia or fear may even apply to scientists and engineers in a university system.

The scientist or engineer in a government laboratory may fall between the two extremes of academia and industry. In fact, it is only now, after years of distrust, that academia and industry are forming alliances for chemical research with trust between the two parties (Mullin, 2013). Furthermore, some government laboratories focus on projects that are top secret, and information on the project cannot be divulged to anyone. Others may work on projects that fall into the public domain, and presentations can be made at any and all scientific or engineering symposia—if it has not already been divulged to all and sundry by means of the various electronic media. But there are always government scientists or engineers who thrive on secretive behavior and do not even divulge information on public-domain projects—according to Sir Francis Bacon (1561–1626; English author, courtier, and philosopher) knowledge is power.

### 6.3.4 TEAMWORK

From an anthropological standpoint, people are an independent branch of the animal kingdom and the members of that branch always have individual opinions and independent methods of performing a task to reach a goal. This is the way that the human brain is guided by the intertwined helical strands of the DNA molecule. Except for a very small percentage of that molecule, sharing and collaboration with others is not exactly programmed inside each and every person—it is evident that each individual is predominantly concerned with his/her rewards as well as the need for power over others, irrespective of the capabilities (or lack of capabilities) of that person. This is one aspect of the Peter Principle in action (Peter and Hull, 1969; Lazear, 2000; Pluchino et al., 2010).

The concept of teamwork is intended to align the team members in a cooperative and usually selfless manner toward a specific business purpose, and it involves sacrifices, sharing of rewards, sharing of blame and punishments, true uniformity, suppression of personal opinions, and so on, which is not very palatable to many. Thus, teamwork involves working with others in a group toward a common goal. This requires cooperating with others, being responsive to others' ideas, taking a collaborative approach to learning, and taking responsibility for developing and achieving group goals. Teamwork uses the background skills of collaboration, mentoring, decision making, and delegation. Commercial organizations will go to extreme efforts to coordinate team-building events in an attempt to get people to work as a team rather than as individuals. Universities are less conscious of teamwork where every professor is his/her own island, having much authority but being willing to accept little responsibility for any of their actions.

In a team, the *principal investigator* (PI) is charged with either conducting research activities on his/her own or supervising those who do. In reality, few scientists and engineers run their own studies. Instead, they hire qualified technicians or laboratory assistants to conduct basic experimental procedures, study the data, perform data assessments, and keep accurate records of the laboratory activities. Unfortunately, in many cases, the regulations or conventions do not require that the principal investigator has any specific training or expertise, other than that of a current/former investigator in the laboratory with (hopefully) expertise in the area under study. Therefore, the extent to which study procedures are delegated and the level of experience, training, and education of those to whom the tasks are delegated are left to the discretion of the principal investigator.

In addition, principal investigators frequently designate subinvestigators. Most often, subinvestigators are chosen from within the department where the research is being carried out, although there may be no requirement that the subinvestigators have specific training or expertise. These individuals are appointed to act as a surrogate for the principal investigator, and they work with the remainder of the project team just as the principal investigator would.

The picture of the scientist or engineer as a loner working in a laboratory is no longer the truth. Consider Dr. Henry Frankenstein (played by Colin Clive) with his faithful assistant Fritz (played by Dwight Frye) looking on yelling "It's alive! It's alive!" as the monster (played by Boris Karloff) moved and twitched on the table during the movie version of *Frankenstein* (released November 21, 1931 and based on the 1818 novel by Mary Shelley). This is no longer the case for scientists and engineers. They do not work alone and are typically members of teams of three or more persons. In fact, one of the most rewarding aspects of being a scientist or engineer is being able to work with other scientists and engineers. Sharing conversation and ideas with other (trustworthy) colleagues to create initiate and develop scientific and engineering concepts is fruitful and rewarding.

Assuming that colleagues are trustworthy and no members of the team suffer from extreme paranoia related to the work habits of their colleagues, there are many ways that teamwork can make science and engineering work better. By working with colleagues, each member of the team can use the collective knowledge and strengths of the other team members. For example, very capable hands-on experimentalists— who work well in the laboratory with the instruments and the reactions—would do well to be partnered with one or more excellent theoretical scientists or engineers who have the talent to model complex phenomena. As a team, they could progress far more effectively than the lone scientist or engineer (the success of Dr. Frankenstein notwithstanding) and the team could make greater strides into the unknown than any individual team member could do alone.

This type of relationship happens all the time in research. The complex questions that scientists and engineers strive to answer give many opportunities for scientists and engineers or people of different backgrounds to contribute to the answers by working together (Sawyer, 2007). For example, Thomas Edison, in order to invent the lightbulb, had many scientists and engineers working for him at the time and built upon existing knowledge. In fact, the first incandescent light was created by Humphry Davy in 1809, followed by contributions from at least 20 scientists who attempted to make a practical incandescent bulb before Edison's team became involved. Edison even purchased a patent from Henry Woodward and Matthew Evans in 1879, the same year he applied for a patent on the lightbulb his team invented.

Another reason that teamwork in science and engineering is the key to innovation and discovery is that it is impossible to keep up with the incredible speed of the advancement of scientific and engineering knowledge. Hundreds of thousands of scientific and engineering papers are published each year. Even this age of a computer in every home and a computer on every desk, one person cannot keep up with all of the old and new information. Teamwork means that each member of the team does not need to know all of the information about the topic under investigation. Each team member focuses on his/her area of scholarship and each member contributes knowledge to solve the problem. The more the merrier—providing the individual members are congenial with each other and work toward a common goal.

In many instances a barrier to effective teamwork is the concept of *tenure* (Chapter 1), which can serve (erroneously) to protect faculty members who do not wish to participate in teaching, research, departmental meetings, and any other activities that might be advantageous to the department, the college, and the university.

Tenure is, in essence, the safety umbrella that is used (incorrectly) to protect faculty members from any form of reprimand and even from dismissal. The system is vehemently protected by university administrators (read: *university bosses*) because that is the system that raised them to the exalted pinnacle of administrator. In addition, it is often the case that many members of the university board of trustees (board of governors, board of regents) do not fully understand the concept of tenure, and the board members should be educated accordingly in order that they (as an entity) may effect good governance (Salmi and Helms, 2013). This situation occurs despite the fact that many of the members of such a board are self-made men and women from the private sector who would not tolerate tenure in their own organizations!

Briefly, a professor who has received tenure means that s/he has been given a lifetime appointment until retirement, except for dismissal for good cause (also known as dismissal with due cause). A common justification for the existence of such an unusually privileged position is the principle of *academic freedom*, which holds that it is beneficial for state, society, and academia in the long run if scholars are free to examine, hold, and advance controversial views without fear of dismissal from their appointed positions (Robinson, 2013).

There is the strong argument that modern tenure systems actually diminish academic freedom, forcing those seeking tenured positions to profess conformance to the same political and academic views as those awarding a tenured professorship (Robinson, 2013). In fact, it is practically career suicide for a young scientist or engineer to hold different scientific, engineering, or academic views to those in a position to award him/her tenure. This may be even more so now that many universities require several years in non–tenure-track positions (such as assistant professorship or postdoctoral Fellowship) before beginning the 5-6-year process preceding tenure. In other words, the system must be protected at all costs!

Another argument against tenure is that professors underperforming in research or teaching cannot be terminated—"dismissal for good cause" is rarely exercised—so typical performance-oriented management techniques from the business world such as reviews, audits, and incentive-based salaries (evaluation) (Chapter 7) are the only tools available—the threat of dismissal without good cause does not (or should not) exist. Nevertheless, many tenured faculty members are expected to (and many do) obtain research funding; to consider this another way, some do but many do not.

And so, the main objection to tenure by many outsiders is that conferring tenure upon a faculty members opens the door to variants of behavior for which the tenured faculty member cannot be (or is unlikely to be) reprimanded, short of being found *in flagrante delicto* in the president's office one dark and stormy night.

### 6.3.5 MENTORING AND OTHER SKILLS

At school, the objective is for the student to learn the fundamental aspects of various subjects in an educational manner and, to accomplish this, the student is subject to the teaching methods of a variety of teachers. At university, the educational process takes on a different format and, consequently, the relationship between student and teacher takes (or, at least, it should take) a different approach. At this time, the student may find the student–professor relationship different to the former student–teacher relationship and the professor should, in addition to teaching, also act as a *mentor* to the student (Bird, 1994, 2001).

Mentoring (Chapter 2) involves taking on the role of a trusted advisor and helper with experience in a particular field. In fairness to many university professors, this is a difficult task and they are not equipped to be mentors, having been poorly mentored during their time as students or never having been mentored at all. Others professors fall into the mentoring role quite easily, actively supporting and guiding students to develop knowledge and experience, or to achieve career or personal goals. In some universities it is the policy to also use senior students (final-year baccalaureate students) as well as graduate students as mentors to help first-year students to adjust to university life. In a commercial setting, if mentoring is used at all, it is likely to be a senior scientist or senior engineer who takes on (rather than is given) the role of mentor to a young scientist or engineer to explain the various aspects of project teamwork and the importance of meeting a project deadline. Whatever the method, the mentoring relationship may be formal or informal but must involve trust, mutual respect, and commitment as the mentor and the student work together to achieve specific goal.

As part of the mentoring program, it is essential that assistance be given so that the neophyte understands the meaning of teamwork (also called *group work*), decision making, delegation, collaboration, network building, and motivation of others.

Teamwork is (1) any activity in which scientists and engineers work together; (2) any activity that has been specifically designed so that scientists and engineers work in pairs or groups, and may be assessed as a group (referred to as formal group work); or (3) when science and/or engineering students come together naturally to help each other with their work, often referred to as informal group work, or take part in any peer group activity such as activity in laboratory classes, tutorials, and teaching.

Decision making involves identifying appropriate evidence and weighing up that evidence to make a choice—such as gathering and assessing information to find the best way to perform an experiment. It also involves taking responsibility for a decision and its outcomes—such as choosing a topic for a group presentation from a number of suggestions.

*Delegation* involves taking responsibility for determining when to ask someone else to make a decision or carry out a task—such as assigning a fair distribution of the workload in a group project, and sharing responsibility with others. For those in the group with the authority to make decisions, it also involves distributing responsibility and authority by giving someone else the discretion to make those decisions—such as the chosen leader of a lab experiment team assigning tasks and decisions to different group members.

Collaboration involves working cooperatively and productively with other team members to contribute to the outcomes of the team's work—such as dividing the workload and sharing the results of the leader's own work with others in the group, or assisting members of the group who are having difficulty completing their tasks.

Network building involves creating contacts with other people and maintaining those contacts—such as meeting someone at a seminar with similar interests, and exchanging contact information with him/her/them. It also involves acquiring and maintaining information about people who might be useful contacts for specific purposes—such as seeking out people established in an industry you hope to work in at some future time. In addition, using a contact in an ethical manner to help meet specific goals and collaborating on projects of importance are also included under network building.

Motivation of others involves the generation of enthusiasm and energy by being positive and focusing on finding solutions and maintaining a positive attitude even when things are not going well. It also includes (1) encouraging others to come up with solutions, listening carefully to their ideas and offering constructive feedback; and (2) being prepared to support others in taking agreed, calculated risks, and not blaming others when things go wrong.

The notion of *mentoring* and the associated issues of group work, decision making, delegation, collaboration, network building, and motivation of others are not new but all should be explained by the advisor/mentor during the time in which the student is in university and not left until the graduate enters the work place. While different companies may place different slants on the various aspects of teamwork cited above, the overall principle will be similar.

Historically, the concept of mentoring is not new and has its origins in ancient times. The original *Mentor* was described by Homer as the wise and trusted counselor whom Odysseus left in charge of his household during his travels. Athena, in the guise of Mentor, became the guardian and teacher of Telemachus, the son of Odysseus. In modern times, the concept of mentoring has found application in virtually every forum of learning. In academia, the term mentor is often used synonymously with *faculty advisor*. However, the fundamental difference between a mentor and an advisor is that in the faculty-student relationship, the mentor is also available to assist the student with personal issues, whereas the advisor typically focuses on course work and other university-related issues. An advisor might or might not be a mentor, depending on the quality of the relationship.

Thus, a mentor is a person who takes a special interest in helping another person (i.e., a student, neophyte scientist, or engineer in an industrial setting) to develop into a successful professional (Noe, 1988; Bird, 1994, 2011; NAS, 1997). Some students, particularly those who move to large commercial laboratories and institutions, find it difficult to develop a close relationship with their laboratory director. They might have to find their mentor elsewhere—perhaps an older co-worker, a friend who is wise in the ways of the world, or another person with experience who offers continuing guidance and support. In the realm of university science and engineering, where mentoring should commence but often does not, a good mentor should (1) seek to help a student optimize an educational experience, (2) assist the student's socialization into a disciplinary culture, and (3) help the student find suitable employment. These obligations can extend well beyond formal university teaching and continue into or throughout the career of the scientist and engineer so that s/he can, in later years, mentor other young first-year scientists and engineers.

Furthermore, an effective mentoring relationship (for scientific and engineering students) must involve mutual respect, trust, understanding, and empathy. A good mentor will share experiences and wisdom as well as technical expertise, and should be a good listener, good observer, and good problem solver. S/he (the mentor) will make an effort to know, accept, and respect the goals and interests of the student. In the end, the mentor should establish an environment in which the student's accomplishments are limited only by the extent of his/her imagination and talent (NAS, 1997).

Finally, it is very presumptive and also incorrect to assume that postdoctoral scientists and engineers require little or no mentoring because they have more experience than undergraduate or graduate students. That might not be true for postdoctoral students, any more than it is for junior faculty. In fact, postdoctoral students, who might have scant supervision, ill-defined goals, and poor access to a community of peers, tend to incur a greater risk of isolation and stagnation than graduate students. A good mentoring relationship can be crucial to the success of postdoctoral scientists and engineers as they develop original research ideas and move toward greater professional independence.

## 6.4 WRITTEN COMMUNICATION

Written communication is one of the methods by which project data shows whether or not the project is worthwhile. However, there are two hindrances to the timely communication of project results to the sponsor.

In the first case, most scientists have met the chemist who only needs 10 (or more experiments) to produce a *possible maybe* as the answer to the problem. So there is no need to use the talents of scientists in the same or different subdisciplines of chemistry (or other science) or related engineering subdisciplines to discuss and solve the problem. In the second case, there is the engineer who has been taught (by others or by himself) that his/her thinking is so logical that whatever s/he writes on paper must be correct and cannot (logic not withstanding) be incorrect. So there is no need to use the talents of engineers in the same or other subdisciplines of engineering or to use scientists in the relevant subdisciplines to discuss and solve the problem. Writing reports should not be held up by either of the example presented above. It is the job of the project manager to see that such cases do not happen.

Scientific and engineering writing, the errant scientist or engineer notwithstanding, is a necessary form of communication and is a style of writing used in fields as diverse as gene divergence and process engineering. Scientists and engineers explain their technology and related ideas to technical and nontechnical audiences. Speaking, as well as writing, is also important in science and engineering. Throughout the career of a scientist or engineer, s/he will confront many writing situations, including proposals, formal reports, and journal articles.

Written communication encompasses a varied collection of methods that scientists and engineers use to document scientific or engineering processes or products—in most cases, the collection is very dependent upon the individual scientist or engineer. Information may be communicated in the form of paper documents, digitally stored text, audio, video, web pages, and a variety of other media including the recent focus on cloud storage. The goal of the practice is to create easily accessible information that will be read and understood by (for the most part) a scientific or engineering audience. For a scientist or engineer to be able to say or write exactly what is on his/her mind is more than a major hurdle and requires considerable thought and effort as well as a conviction about what is being written. It should always be kept in mind that even though the facts have been clearly communicated, that does not mean that the individual reader or the collective reader (the audience) will necessarily regard them as important.

The value of such communications is that scientists and engineers (1) make information more useable and accessible to those who need that information, and (2) the goals of the organization or companies are advanced. Indeed, scientific and engineering writing and publication of data and other forms of communication signifies the endpoint of a research project that has been performed, completed, and peer reviewed and accepted, and that complements teaching. Although in many cases, some projects may be ongoing and never end, the completion of each phase must be communicated to the relevant authorities.

The need to disseminate scientific and engineering knowledge and expertise led to the birth of writing in scientific journals (Shaw and Despota, 2013). The scientific publications that resulted provided additional benefits such reputation among peers and monetary benefits, beyond spreading knowledge. Interestingly, publishing scientific content in journals regularly is often a prerequisite for appointment or promotion in several institutions across the globe. Thus, with scientific publications becoming synonymous with job survival, scientists and engineers must have a publishing record built up during their respective careers. The emphasis is not always on publication of data—the data may be proprietary to a company or a client (if the data were assembled by a paid consultant) but at some time will (should) appear in the form of a patent.

Writing technical papers for publication or a technical report, as a form of communication, has numerous benefits, and one of the most important of these is the training undertaken by the young scientist or engineer to evaluate the published work of others. Effective writing is an important component of the careers of all scientists and engineers and should be cultivated at an early stage of their respective careers (Hayes and Flower, 1980; Paradis and Zimmerman, 1997; Peh, 2007; Dawson and Gregory, 2009). Furthermore, a successful publishing career involves writing technical papers for a highly specific scientific or engineering audience, and it may take many authors years to develop a writing style that results in a high percentage of well-accepted papers. To acquire such writing skills the scientist or engineer can work alone, in isolation from colleagues, and hope to learn from rejection letters and from harsh peer reviews, or s/he can build an informal team of fellow scientists and engineers who are both critical and supportive and who will read and comment constructively (another form of mentoring) on any written work.

If the scientist or engineer is new to an organization or academic department, it necessary that s/he wishes to quickly determine who will be supportive of his/her aims versus who may be less than helpful. The *coffee klatch* group who tend to be the department or company gossips may be of little value but a novice author can learn much from established authors by passing drafts to them for assessment and to get their recommendations for getting published. However, it is wise and a common courtesy to discuss the possibility of his/her review of the paper/report with the established author rather than spring the written work on him/her without a heads up.

Receiving constructive comments from an established author will continue the tradition in science and engineering that sees the writing process as a series of decisions and choices, but this tradition can only be asserted if the writer is prepared to answer a number of questions about the work once it is published (as in the case of a scientific or engineering paper) or read (as in the case of a project report). If the established author merely asks "Why do you wish to publish this work?" while indicting his very untidy, dusty desk piled high with papers and adding "I have six (or some number of) papers/reports here," it is advisable not to use this person as a reviewer for future papers/reports. The young scientist and engineer can be guaranteed that the comments from such a person will be far from constructive and lean toward destructive comments. This person will also be an abject (in the sense *extremely bad*) failure as a mentor.

To most scientists and engineers it may seem reasonable to suppose that when they put thoughts on paper, it may not always be clear to the reader how these thoughts might interact, especially in a coordinated manner. The best way to move ahead with writing is for the would-be writer to understand the nature of the thoughts (and choices) that make some forms of writing good and others bad, although the ultimate issue is the difference between the two forms of writing.

One of the critical aspects of the scientific and engineering progress is, for example, the reporting of new results in technical journals in order to disseminate that information to the larger community of scientists and engineers. This contributes to the dissemination of knowledge within a scientific or engineering discipline and very often provides information that helps other scientists and engineers interpret their own experimental results (Shaw and Despota, 2013). If the data are proprietary to a company, the dissemination of the knowledge in a report for company-wide circulation can have the same result. Most journals accept papers for publication only after peer review by scientists and/or engineers who work in the same field and who recommend the paper be published (usually with some revision) or rejected (for various reasons). The same is often true for company reports—the report is peer reviewed within the same department as the author before dissemination on a company-wide basis.

No single course of action can prepare the scientist or engineer for every communication situation that s/he will face. Nevertheless, s/he should be able to handle most situations if there is a preliminary consideration of any constraints. One of these constraints is format, and it is necessary to understand that there is no universal format for scientific and engineering writing. The formats used in one organization are not the same ones that scientists and engineers use in a different organization. Anyone who has read scientific and/or engineering papers will have noticed that a standard format is frequently used and this allows (and encourages) a scientist or engineer to present his/her information clearly and concisely.

Regardless of whether a scientists or engineer is writing a technical paper for publication in a journal, a technical report for company distribution, a collection of slides for presentation at a meeting, an e-mail, or a memorandum, the communication must represent the abilities of the writer as well as the character of the writer

(Medawar, 1979; Day, 1994; Markel, 1996; Alley, 1996; Alley et al., 2006). Using an appropriate tone is essential and the language must be considered very carefully so that the writer does not come across as arrogant, overconfident, or too demanding. In all correspondence, the writer should focus on being concise and accurate. Key points should be presented early in all documents so that those same points stand out from the rest of the text.

In addition, the document must appeal to the designated audience and the writer must ensure that the tone is appropriate for the readership. The writer must be respectful and polite to his/her readers and there must be sufficient information about the problem under investigation. Indeed, there must be enough information in the document for the reader to understand the context of the problem. Addressing such issues will ensure that the written communication helps the scientist and engineer to build and maintain a professional relationship with his/her colleagues and peers.

Finally, before a scientist or engineer commits words to paper in a scientific or engineering document, s/he should understand the subject matter that is being communicated to the reader or the audience. In other words, pages should not be filled with (often meaningless) sentences unless there is a general (even fixed) idea where those sentences are headed. Even after there is a general understanding of the subject matter, the scientist or engineer should not begin writing until the various writing constraints (which are those aspects of the writing that the writer does not control) have been analyzed. These constraints include (1) the audience for the document and (2) the format of the document. Another constraint, not discussed here, is mechanics, which comprises grammar, punctuation, and usage (Chapters 5 and 6). Besides the constraints of audience and format, there is also the writing style (Strunk, 1918; Strunk and White, 1959; Flower and Hayes, 1981) to be considered, but this is an aspect of the writing that the scientist or engineer does control.

More specifically, the organizing process allows the writer to identify categories, to search for subordinate ideas that develop a current topic, and to search for superordinate ideas which include or subsume the current topic. At another level the process of organizing also attends to more strictly textual decisions about the presentation and ordering of the text. That is, writers identify first or last topics, important ideas, and presentation patterns.

Finally, writing is the most important means for communicating scientific and engineering work. Research and publication complement teaching. There are many reasons for writing, one of the most important of which is to better appreciate and evaluate the published work of other scientists and engineers.

## 6.5   THEORETICAL KNOWLEDGE TO PRACTICAL EXPERIENCE

The road from theoretical knowledge to practical experience is often fraught with bumps, ridges, and potholes and is very rarely smooth. Unfortunately, other than the ethical aspects of traveling the unsmooth road, very little can be taught about this in universities and the student, for the most part, has to rely on practical experience and (if s/he is lucky) objective mentoring.

*Theoretical knowledge* and *practical experience* can be defined very simply as *thinking* and being *hands-on*—without any disrespect to the scientist or engineer who is a theoretician or a practitioner. In modern science and engineering, the term *theory* refers to scientific theories and engineering theories, a well-confirmed type of explanation of a phenomenon, made in a way consistent with the investigative method and fulfilling the criteria required by modern science and engineering. Such theories are described in such a way that any scientist in the field is in a position to understand and either provide empirical support (*verify*) or empirically contradict (*falsify*) it. Scientific and engineering theories are the most reliable, rigorous, and comprehensive form of scientific and engineering knowledge, in contrast to more common uses of the word theory that imply that something is unproven or speculative (which is better defined by the word *hypothesis*). In addition, the development of the scientific and engineering method (*investigative method*) of inquiry has made a significant contribution to how knowledge is acquired and is based on gathering observable and measurable evidence subject to specific principles of reasoning and experimentation.

Scientific and engineering knowledge may not involve a claim to certainty and maintaining a healthy curiosity (some would say skepticism) means that a scientist or engineer will never be absolutely certain when the data (or hypotheses) are correct and when they are not. It is thus an irony of the investigative method that a scientist or engineer must doubt the data even when correct, in the hopes that this practice will lead toward greater certainty. Indeed, as a part of the education process, the scientist and engineer must learn the role of anomalous data (Chinn and Brewer, 1993). It is in the area of anomalous data use that the scientist and engineer (of any age and experience) can continue to learn their greatest lesson.

Anomalous data in science and engineering refer to data (or information) that seems to contradict scientific or engineering ideas (or theories) about a particular phenomenon. The key to using analogous data here is that scientists and engineers must have used their previous knowledge (the concepts or theories that they proposed) to make predictions and not promote naive theories or wild ideas. When anomalous data are the result of experimentation, the scientist or engineer can observe the contradiction between the prediction and the outcome. The next step is to confirm the anomalous data as being true to the experimental method and (if necessary) to revise the explanation to accommodate the experimental results. The data must not be rejected out of hand because they do not fit the theory.

It is at this stage that many scientists and engineers decide that the experimental design was not incorrect, or that the failed hypothesis was not incorrect, and they push forward to explain the experimental results. If the conduct of research is not monitored closely by peers and supervisors/mentors, a situation can exist where bending of the truth (it may not be called cheating but that is what it is) occurs and the objectivity of the researchers is lost. When this happens, integrity is forfeited (Speight and Foote, 2011; Anderson and Kamata, 2013; Bretag, 2013).

For example, the experiment that failed becomes the experiment that succeeded because of a datum point (or data points) that has just been discovered. The defeated hypothesis becomes the successful hypothesis because the experimental design produced a datum point that the researcher was seeking. The means by which the datum point came about is another issue and is looked upon as good fortune by the supposedly unbiased and totally honorable researcher involved. Or the datum point was discovered in a blinding flash of untruthful inspiration by coworkers of the researcher who knew how important such a datum point would be. The experiment that failed becomes the experiment that provided crucial proof of a concept.

On the other hand, too many points can be a hindrance to a researcher and lead to hours (or minutes or seconds) of heart-rending consideration. The result might be that out of 24 shotgun-patterned points on an x–y chart, 18 points are omitted as flyers. The result is an x–y relationship on the chart that gives credence, even proof, to the hypothesis and results in wide acceptance of the hypothesis and copious accolades for the researcher. After the success of such a brilliant hypothesis, there are few if any who will repeat the work to determine if the data are correct (perhaps because of funding constraints). The hypothesis lives on, and it is only after serious issues have been raised at some future time that the hypothesis is reworked. By then the original researcher may have retired after a distinguished career with a reputation that is beyond reproach. Younger researchers who could not make any sense of the hypothesis and report data that contradict the data of the distinguished researcher are, at first, criticized and ostracized until the truth become known.

Deleting anomalous data points is hardly uncommon—initially, all of the data points are printed on a scatter plot, and so-called flyers that do not match the plot are omitted. This is such a familiar situation in research that there are many reasons for deleting the nonconforming data. This would imply that there are certain situations in which the practice is considered to be acceptable, but such deletion actually amounts to misrepresentation. Flyers can be influential or not influential insofar as they are far removed from and inconsistent with the rest of the data or are far removed from but consistent with the rest of the data. In the former case, summarization and analysis of the data, both with and without the outliers, can be performed and the different inferences and conclusions are assessed—with and without the flyers. Nevertheless and in either case, all outliers must be reported. To do otherwise is tantamount to technical fraud through dishonesty (intentional deception) (Resnik, 1998, 2000; Speight and Foote, 2011; Anderson and Kamata, 2013; Bretag, 2013).

Omitting or ignoring anomalous data is a reprehensible act that dishonors scientists and engineers (from whom we expect the truth; anyone, scientist or not, can lie and deceive). The occurrence of anomalous data provides the researcher with an opportunity for both instruction and learning—anomalous data should be given consideration rather than being rejected out of hand because (1) the experimentalist receives feedback on the experimental procedure and (2) the experimentalist receives feedback on the theoretical deductions. By analyzing the data *in toto* the scientist or engineer can have a better understanding of the application of the theory and make the necessary changes.

## 6.6  INTELLECTUAL PROPERTY AND ITS ROLE IN EDUCATION

One aspect of educating scientists and engineers that is often lacking in institutes of learning is the teaching or discussion of the fundamentals of intellectual property rights. Such discussions are well-covered by companies in the form of intracompany workshops and lectures for new employees of all disciplines.

Intellectual property is a legal field that refers to innovations that are protected from unauthorized use by patents, trademarks, copyrights, and related rights. Under intellectual-property law, the holder of a patent has specified exclusive rights to the invention (WIPO, 2000; Moore, 2004). One category of intellectual property is collectively known as *industrial properties*, which are typically created and used for industrial or commercial purposes. A patent may be granted for a new, useful, and nonobvious invention, and gives the patent holder a right to prevent others from practicing the invention without a license from the inventor for a certain period of time. A trademark is a distinctive sign that is used to prevent confusion among products in the marketplace. A trade secret is nonpublic information concerning the commercial practices or proprietary knowledge of a business.

Intellectual property, in the form of patents, protects an invention and the rights of the inventor and/or the patent owner. Patents provide inventors or those deriving title from them the right to prevent others from making, selling, distributing, importing or using their invention, without license or authorization, for a fixed period, normally in the order of 20 years from the application date. Patents are subject to an examination by the Patent Office before being granted and to the payment of renewal fees thereafter. In return, the applicant for the patent is required to disclose the invention in the *patent specification* and to define the scope of the patented invention in *patent claims*.

There are three further requirements for an invention to be patentable: (1) novelty, normally over anything disclosed publicly anywhere; (2) inventive step or nonobviousness—the invention would not have been obvious to a person skilled in the art at the time the application for a patent was filed; and (3) industrial applicability. Patents are limited to the country for which they have been granted. Granted patents can be contested in the courts or (sometimes) patent offices in validity proceedings or as a defense to an allegation of patent infringement.

In most countries, novelty is destroyed by public disclosure by any means (oral or written). On the other hand, some countries, including the United States and Japan, allow disclosure to be made without prejudicing a patent application as long as patent application is filed within 3–12 months of the disclosure (the *grace period*). There are in fact many forms, and potential forms, of grace period—in the United States, the process is a *first-to-invent* system rather than a *first-to-file* system and an inventor has the option of producing evidence that s/he made the invention before a prior publication of somebody else. This right leads to so-called *interference* proceedings, in which an interested party has the right to challenge a patent on the grounds that the subject matter had already been invented.

In summary, innovation and intellectual property rights are the life-blood of science and engineering, but most nonindustrial scientists and engineers learn about intellectual property rights after the fact, and often by trial and error. There is the need for course related to the definition and purpose of intellectual property right in universities so that young scientists and engineers can be prepared for what may lie ahead.

# REFERENCES

Alley, M. 1996. *The Craft of Scientific Writing*, 3rd edn. Springer, New York.
Alley, M., Schreiber, M.M., Ramsdell, K., and Muffo, J. 2006. How the design of headlines in presentation slides affects audience retention. *Technical Communication*, 53(2): 225–234.

Anderson, M.S. and Kamata, T. 2013. Scientific research integrity as a matter of transparency. In: *Global Corruption Report: Education* (Chapter 3.19). Routledge, Taylor & Francis Group, Abingdon, Oxford.

Bird, S.J. 1994. Overlooked aspects in the education of science professionals: Mentoring, ethics, and professional responsibility. *Journal of Science Education and Technology*, 3: 49–55.

Bird, S.J. 2001. Mentors, advisors and supervisors: Their role in teaching responsible research conduct. *Science and Engineering Ethics*, 7: 455–468.

Bretag, T. 2013. Short-cut students: From academic misconduct to academic integrity. In: *Global Corruption Report: Education* (Chapter 3.13). Routledge, Taylor & Francis Group, Abingdon, Oxford.

Chinn, C.A. and Brewer, W.F. 1993. The role of anomalous data in knowledge acquisition: A theoretical framework and implications for science instruction. *Review of Educational Research*, 63(1): 1–49.

Dawson, N.V. and Gregory, F. 2009. Correspondence and coherence in science: A brief historical perspective. *Judgment and Decision Making*, 4(2): 126–133.

Day, R.A. 1994. *How to Write and Publish a Scientific Paper*, pp. 8–14. 4th edn. Oryx Press, Phoenix, AZ.

Flower, L.S., and Hayes, J.R. 1981. A cognitive process theory of writing. *College Composition and Communication*, 32(4): 365–378.

Forbes, R.J. 1958a. *A History of Technology*. Oxford University Press, Oxford.

Forbes, R.J. 1958b. *Studies in Early Petroleum Chemistry*. E.J. Brill, Leiden, The Netherlands.

Forbes, R.J. 1959. *More Studies in Early Petroleum Chemistry*. E.J. Brill, Leiden, The Netherlands.

Forbes, R.J. 1964. *Studies in Ancient Technology*. E.J. Brill, Leiden, The Netherlands.

Griffin, E. 2011. *The Mechanical Age: Technology, Innovation and Industrialization. Short History of the British Industrial Revolution*. Palgrave Macmillan Ltd., Basingstoke, Hampshire, United Kingdom.

Hayes, J.R. and Flower, L.S. 1980. Identifying the organization of writing processes. In L. Gregg and E. Steinberg (eds), *Cognitive Processes in Writing: An Interdisciplinary Approach*. Lawrence Erlbaum Associates, Hillsdale, NJ.

Hazelkorn, E. 2011. *Rankings and the Reshaping of Higher Education: The Battle for World-Class Excellence*. Palgrave MacMillan, Basingstoke, Hampshire, United Kingdom.

Jindal-Snape, D. and Snape, J.B. 2006. Motivation of scientists in a government research institute: Scientists' perceptions and the role of management. *Management Decision*, 44(10): 1325–1343.

Laplante, P.A. 2012. *Technical Writing: A Practical Guide for Scientists and Engineers*. CRC Press, Taylor & Francis Group, Boca Raton, FL.

Lazear, E.P. 2000. *The Peter Principle: Promotions and Declining Productivity*. Revision 10/12/00. Hoover Institution and Graduate School of Business, Stanford University, Stanford, CA.

Markel, M. 1996. *Technical Communication*, pp. 420–421, 4th edn. St. Martin's Press, New York.

Medawar, P.B. 1979. *Advice to a Young Scientist*. Alfred P. Sloan Foundation, Library of Congress, Washington, DC.

Moore, A.D. 2004. *Intellectual Property and Information Control: Philosophic Foundations and Information Control*. Transaction Books, New Brunswick, NJ.

Mullin, R. 2013. University Inc. *Chemical and Engineering News*, 91(31): 12–16.

NAP (National Academies Press). 1995. *Reshaping the Graduate Education of Scientists and Engineers*. National Academies Press, National Academy of Sciences, National Academy of Engineering, Institute of Medicine, Washington, DC.

NAS (National Academy of Sciences). 1997. *Adviser, Teacher, Role Model, Friend: On Being a Mentor to Students in Science and Engineering.* National Academy of Sciences, National Academy of Engineering, Institute of Medicine, National Academy Press, Washington, DC.

Noe, R.A. 1988. An investigation of the determinants of successful assigned mentoring. *Personnel Psychology*, 41: 457–479.

Paradis, J.G. and Zimmerman, M.J. 1997. *The MIT Guide to Science and Engineering Communication.* The MIT Press, Cambridge, MA.

Peh, W.C.G. 2007. Scientific writing and publishing: Its importance to radiologists. *Biomedical Imaging and Intervention Journal*, 3(3): 55.

Peter, L.J. and Hull, R.R. 1969. *The Peter Principle: Why Things Always Go Wrong.* William Morrow and Company, New York.

Pluchino, A., Rapisarda, A., and Garofalo, C. 2010. The Peter principle revisited: A computational study. *Physics A: Statistical Mechanics and its Applications*, 389(3): 467–472.

Resnik, D.B. 1998. *The Ethics of Science: An Introduction.* Routledge Publishers, New York.

Resnik, D.B. 2000. Statistics, ethics, and research: An agenda for education and reform. *Accountability in Research*, 8: 163–188.

Robinson, D. 2013. Corrupting research integrity. In: *Global Corruption Report: Education* (Chapter 3.18). Routledge, Taylor & Francis Group, Abingdon, Oxford.

Salmi, J. and Helms, R.M. 2013. Governance instruments to combat corruption in high education. In: *Global Corruption Report: Education* (Chapter 3.2). Routledge, Taylor & Francis Group, Abingdon, Oxford.

Sawyer, K. 2007. *The Creative Power of Collaboration.* Basic Books, Perseus Books Group, Cambridge, MA.

Shaw, M.M. and Despota, K. 2013. Journals: At the front line of integrity in academic research. In: *Global Corruption Report: Education* (Chapter 3.20). Routledge, Taylor & Francis Group, Abingdon, Oxford.

Shola Omotola, J. 2013. Corruption in the academic career. In: *Global Corruption Report: Education* (Chapter 3.15). Routledge, Taylor & Francis Group, Abingdon, Oxford.

Singer, C., Holmyard, E.J., Hall, A.R., and Williams, T.I. (eds). 1954–1959 and 1978. *A History of Technology*, vol. 1–7. Clarendon Press, Oxford. Note: Volume 6 and 7, 1978 Edition, Editor: T.I. Williams.

Speight, J.G. and Foote, R. 2011. *Ethics in Science and Engineering.* Scrivener Publishing, Salem, MA.

Strunk, W. Jr., and White E.B. 1959. *The Elements of Style.* Macmillan, London.

Walters, D.E. and Walters, G.C. 2011. *Scientists Must Speak*, 2nd edn. CRC Press, Taylor & Francis Group, Boca Raton, FL.

WIPO (World Intellectual Property Organization). 2000. *Guide to Intellectual Property Worldwide Second Edition.* Publication 479(E). World Intellectual Property Organization, Geneva, Switzerland.

# 7 The Future

## 7.1 INTRODUCTION

In the past, the system of graduate education in science and engineering in the United States set the international standard, especially in preparing students to work successfully at the cutting edge of research. This must continue but with changes that are sufficient to encourage the students to perform, and these changes must start at the school level (NRC, 1997; Felder et al., 2000; Osborne, 2007). In addition, graduate schools also must fill the growing need for advanced expertise in science and engineering. Until these two aspects of education in science and engineering are satisfied, there will always be the need for substantial improvement in the education of scientists and engineers.

## 7.2 THE FUTURE OF EDUCATION

Education in the scientific and engineering disciplines is very important for innovation and development. The current (and continuing) competitive environment demands that students at all levels have in-depth and practical knowledge and understanding of their specific area along with interpersonal and communication skills (NSF, 1996). Indeed, during the past several decades there have been numerous studies conducted by organizations such as the National Academies, federal agencies, and professional societies suggesting the need for new paradigms in science and engineering education to better address the needs of the twenty-first century, which will soon be moving (believe it or not) toward the twenty-second century!

In fact, science and engineering in the twenty-first century will explore new concepts and develop these concepts to commercialization. However, much of the science and engineering that is likely to evolve may not fall within traditional disciplines but new subdisciplines that are concerned with understanding and designing complex scientific and engineering systems. New challenges will catalyze technology development in science and engineering, and to do this many educational institutions will need to create multidisciplinary and multidepartmental graduate programs in science and engineering to accommodate the globalization of science and engineering education (Burland et al., 2004; NAS, 2005). There will also be strong economic pressure for the graduate student to complete coursework quickly and to move into a laboratory, where, typically, s/he will work on a specific project with the need to produce results to ensure renewal of the funding as well as the freedom to explore novel questions and other educational aspects (Sung et al., 2003; Osborne, 2007; King, 2011).

Indeed, in the twenty-first century, universities could be faced with serious problems such as the possible conflicts between teaching and research and the effectiveness of both areas, and will need to remove such a predicament. The university must focus equally on teaching and research in order to meet the globalization of science and engineering and must work to continually serve as the prime location for the delivery (teaching) and production (research) of knowledge. In order to accomplish these goals, educational organizations (starting in schools and not excluding universities) will need to give close (perhaps the word should be "closer" or "diligent") attention to (in the context of this book) three major areas: (1) the student body, (2) information and guidance through meaningful and helpful mentoring, and (3) the structure of graduate schools.

### 7.2.1 THE STUDENT BODY

There is a substantial variation in the composition of the student body in science and engineering by sex, race, and ethnicity of degree recipients (NSF, 2012). To some observers, such a variation may be a nonissue, especially with male (nonfemale and nonminority) students, but it is often a major issue with female and minority students, especially since these students are projected to occupy a larger proportion of the student body in the twenty-first century (Matthews, 2007). An effort must be made to determine whether there are (unseen and unspoken) barriers to the entry of female and minority students into science and engineering education, including issues perceived as barriers by members of the two groups, and efforts should be made to remove any such barriers—perceived or real (Cockburn, 1985; Etzkowitz et al., 1994; Cuny and Aspray, 2000; Conefrey, 2001; Shaywitz and Hahm, 2004; Hall, 2007). Moreover, actions to encourage increased participation should be devised and implemented.

The presence of women and minority-group students in the United States is still small relative to the university population as a whole in nearly all science and engineering fields, and many often leave the profession for a variety of reasons (Sonnert, 1990; Kvande and Rasmussen, 1994; Preston, 1994; Kvande, 1999; Hersh, 2000; Morgan, 2000; Hathaway et al., 2001; Washburn and Miller, 2005; Gill et al., 2010; Hunt, 2010; Singh et al., 2013). In the long run, it is in the interest of all to recruit the most able students into science and engineering, irrespective of ethnic background and, especially, gender (Hacker, 1989; McIlwee and Robinson, 1992; Lewis, 1993; Frehill, 1997; Faulkner, 2000; Lackland and De Lisi, 2001; Jacobs and Winslow, 2004). In short, there must be continuing action to ensure that all people with the necessary scholastic talent have an equal opportunity to enter science and engineering careers.

It is quite likely that a *critical mass* of female and/or minority students is particularly important for members of the female and minority groups who, as single students in a group, may often have the disadvantage of feeling the *one and only* syndrome. Females still comprise a minority of the employed scientific and engineering labor force in academia and in the industrial (nonacademic) environment (NAP, 1994; ASEE, 2009; IChemE, 2013). On the other hand, female students and minority students are forming an increasing proportion of the school and university populations and will continue to do so for the foreseeable future (McBay, 1989; Etzkowitz et al., 1994; Anderson and Kim, 2006; Fiegener and Proudfoot, 2013). Furthermore, many of these students may lean toward a career in science and engineering, which could require (dictate) that teachers/professors use texts, lesson plans, and other resources to reduce the mismatch in the relative numbers and develop curricula more suitable to a mixed student body (Tapia, 2009).

The key for universities and any institutions of higher education is *not* to play a "numbers game" to show the proportion of female students and minority students in the university student body but to provide academic advice and financial-aid options that encourage students to enroll full time and reduce their need to work (outside of the university) for more than 14 h a week (King, 2002). In short, education in the scientific and engineering discipline should be available to all students with the necessary scholastic aptitude rather than to those students who (in reality, those students whose parents) can afford the tuition fees. And there is always the question as to whether or not the SAT is a fair test (Lewin, 2013; Balf, 2014) and, once into university, whether or not the education received is worth the money spent in fees (Getz and Siegfried, 2010; Reynolds, 2014a).

In addition, in the midst of the large influx of female students, universities must take steps to protect female students from sexual assault (although sometimes a male student may be assaulted by females) (Winerip, 2014). Moreover, the victim should not be made to feel guilty and must be given an appropriate audience to express herself and present her case.

Finally, an important aspect of the student body relates to the issues of cheating (for whatever reason) in universities. In spite of whatever protestation may arise from such a statement, it must be recognized that cheating occurs—in academia. Dishonesty violates all educational protocols by giving some students an unfair advantage. But it does not stop with the students (Chapter 2) (Speight and Foote, 2011). Steps must be taken to stop this kind of activity, even by the students themselves, who should report cheating by other students and unfair practices (which can be surmised—in many cases as cheating)—by any members of the faculty (Speight and Foote, 2011; Cheung, 2014). Such dishonesty in any form, let alone academic dishonesty, is a serious offense and may well continue beyond graduation (baccalaureate and higher degrees) and well into adult life (Chapter 4).

## 7.2.2  INFORMATION AND GUIDANCE

Advice for students should not be limited to the personal knowledge of the faculty member who serves as the adviser/mentor. Departments—through the faculty members—should understand and be able to convey to the students the various relevant employment prospects (NAS, 1997a). One way to start this passage of information from the faculty to the students is to track—with the assistance of alumni affairs offices—the post-university careers of their own former graduates in a systematic and meaningful manner. Merely parroting a perfect number (or any number for that matter)—such as "all of the graduates from my class have found employment"—is not sufficient. It may be that a number of students from that class had ready-made-return-to jobs after graduation or some may have been recruited by the professor as candidates for graduate degrees. In fact there should be definitive moves by universities to improve all aspects of undergraduate teaching though hiring high-quality teaching faculty and to ensure that one (e.g., research) is not favored over the other (e.g., teaching) (NRC, 2003; Basken, 2011).

Indeed, it is the duty of the professor/adviser/mentor to pass on to his/her graduating scientists and engineers the most up-to-date, accurate, and accessible information so that the students can make informed decisions about professional careers. In these days of computerization, it should not require any great effort to store information related to career choices for graduating students. However, given the diversity of the information for which there is a need, it is should be obvious to the various faculty and to the university administrators (and also professional societies) that the responsibility for providing such information must be shared by all partners in the graduate-education enterprise. Academic departments should provide employment information and career advice to prospective and current students in a timely and informative manner that will assist students to decide on career paths.

In fact, academic departments can focus attention on the importance of career choice and it would be helpful if more departments, in describing their programs to potential students, routinely provided data that is relevant to career choice, such as location of job placements, salaries, and unemployment rates for specific job markets and for the scientific or engineering discipline—such actions may assist in transforming undergraduate education in science and engineering (NAS, 1997b). In addition, a student should be able to consult employment records assembled by the department related to the careers of the entire departmental faculty. Such records will help student(s) to form realistic expectations while, at the same time, preventing unrealistic expectations.

### 7.2.3 Graduate Programs

Once enrolled, a graduate student (or, more likely, the professor/mentor) might find many reasons to select a relatively narrow subject for study. A student might be fascinated by a particular field of scientific or engineering knowledge and foresee specialization as the most certain route to a research position. However, only if the selected field aligns with the research interests of a professor will the student have an educationally enriching chance to work in such a field. The student will also need assurance—which should be given—that the work for the advanced degree will lead to completion of his/her educational experience followed by a professional placement in the academia or the outside world—in short there is a need to reshape, reorganize, and enhance postdoctoral operations (NAP, 1995; NAS, 2000). However, there are disadvantages to overspecialization in graduate school on a very narrow subject area, which, although not immediately apparent, can be very real for the student. Excessive concentration in a particular subfield of science or engineering can limit the later research contributions and can cause the student(s) to suffer from the inability to recognize and enter newly emerging kinds of research.

Indeed, it is difficult to gauge whether a specialty chosen at the commencement of graduate-school work will be desirable in the job market or still be on the exciting forefront of research when the graduate years are over. It is not merely a matter of the professor/mentor having money available for a project but it is really a matter of how that project will help the student in the job market—typically after 4–6 years. But there are those professors who are quite adamant that to allow a doctoral candidate to leave after 4–6 years of work is unacceptable because s/he (the candidate) is not ready, and that a period of 7–10 years is necessary for completion of the work (Chapter 2). This serves only to remove the student further from the realities of the job market, and such practices are of no value whatsoever to the student—things do change in a 7–10 year period! In fact, to produce scientists and engineers who are versatile, professors should provide options that allow students to gain a wider variety of academic and professional career skills. If it takes 7–10 years for a student to complete a doctorate, shame on the professor/mentor.

There is a real need to control the time taken by any student to obtain a degree at any level of the academic degree tree. Such controls should not be construed as the formulation of additional requirements that would monitor the time a student spends in, for example, a graduate program. The steadily lengthening time to degree—and, more importantly, the time to first academic or nonacademic employment—has already been stretched beyond credible limits for a variety of reasons, which appear to favor (1) the university in terms of the collection of additional fees as well as a cut of the research funding for administrative costs, or (2) the professor in terms of the production of additional publishable data made possible by retaining a relatively low-paid assistant.

In short, graduate programs should provide (in a timely manner) options that equip students for a wide variety of career opportunities and should also provide (and promote) the adaptability of the student. In fact, adaptability can be enhanced in ways that allow graduate students to benefit from a wider variety of academic preparation by giving them a grasp of the broad fundamentals of the major scientific or engineering fields as well as familiarity with several. The student should be encouraged to read widely around his/her particular research topic area and gain a wider view of the relevant subdiscipline of science or engineering.

It would also be of benefit if graduate students could participate in off-campus experiences to acquire the skills desired by an increasing number of employers, especially the ability to communicate complex ideas to nonspecialists and the ability to work in teams of interdependent workers. The internship in off-campus settings is one option that needs to be expanded by universities. Project-oriented teams in corporations provide potential opportunities for collaborative interactions and exposure of the student to challenging practical problems. In fact, joint industry-university projects (in which the students serve as interns) should be explored (even further than they are at the moment) as part of the preparation of students for the nonacademic world and the realization that there is (and must be) life after the university (Tuhus-Dubrow, 2013)—for some students, it may even be possible to complete the dissertation work in an off-campus (industrial) setting. Such projects will also acquaint faculty members with the needs and organizational cultures of nonacademic employers. However, the important issue that comes to the fore relates to the access of the intern to company-proprietary information, the unauthorized and premature release of which can cause the student and the university considerable embarrassment.

## 7.3 EVALUATION

The contradiction in many education systems is that, on the one hand, many teachers undergo some form of teacher training, which helps the would-be/prospective/neophyte teacher understand methods of presentation of information and knowledge to students. On the other hand, many professors do not have an inkling of the method of teaching and delivering information and knowledge to students. Brief experience as a teaching assistant (Chapter 2) may be the only training in the professorial repertoire. And then there is *tenure*.

Many universities, filled to capacity by faculty members who were awarded tenure during the 1980s and 1990s, have curtailed hiring in the last two decades—some of the senior faculty have essentially retired on the job but continue to receive the full pay of the professor rank (Chapter 1). Hopefully, an increase in faculty retirements will eventually relieve these pressures on the hiring of new academic faculty. However, in the eyes of many doctoral graduates the attractive pull of an academic career now represents a last resort, and they seek jobs in the worlds of government laboratories and commercial laboratories. The universities have no one to blame for this situation but themselves for propagating a system that led them to this quagmire of too many aged and inactive faculty members.

Awarding tenure on the basis of longevity rather than performance, as is sometimes the case, (Chapter 1) means that nonperformers who have been the happy recipients of tenure cannot go anywhere else—their personnel records are certainly not of the stellar category—and they do not wish to go anywhere else, so they remain at their home base. The opportunities for maximum advancement of such faculty have been attained when the individual has reached the rank of full professor and they settle in for a comfortable life of semiretirement or retirement while being paid for a job not well done or, in some cases, not done at all. The challenge is for faculty members in all university departments to look objectively at their colleagues and determine which one, two, or more colleagues fit this mold and what course of corrective action should be taken.

Corrective action usually should follow an annual (or semiannual) evaluation, but the subject of teacher/professor evaluation (Chapter 1) has been a sore point for many years (if not many decades) and is certainly worthy of further discussion here at the risk of raising the ire of many teachers/professors—as it has a direct bearing on the future of the education system (Marx, 1990). However, it is not the purpose of this section to deal with hours worked as there are several publications that relate to this issue and should be consulted for further information (OECD, 2012 and references cited therein).

The evaluation of the performance of teachers/professors, as in all evaluations of performance, is a difficult but necessary task that must be addressed with seriousness, meaning, and purpose. In many schools and institutes of higher education where tenure is the order of the day, evaluations are decried by teachers/professors and the salary raise that is decided on the basis of the results of the evaluation are condemned even more loudly and openly. Yet, in order to increase the quality of teaching in schools and universities, teachers/professors must be evaluated and the evaluation *must* be based on teacher performance. It is no longer viable to use the excuse of a poor class of students to justify poor teacher/professor performance. Parents are paying fees and a variety of other costs to have their children educated, and in such cases, the clients (parents) have a right to expect a reasonable return (quality teaching) for their investment (Stensaker, 2013). In addition, students paying their own fees (as well as the associated costs) have an equal right to be educated and expect a reasonable return (quality teaching) for their investment. In addition, many students take out loans to fulfill their educational dreams and repayment of the loan(s) can carry over well into professional life (Weingarten, 2013). Furthermore, it is interesting to note that student debt more rapidly accumulates at universities with the highest-paid executive leaders (Lewin, 2014), which points to the need for the reevaluation of the salaries of the highest-paid executives in a university compared to the quality of education in that university.

In addition, the education of foster children is often neglected to the extent that very few obtain a baccalaureate by the time they have reached the age of 24—they are often the forgotten children of the education system (Winerip, 2013). In short, universities need to dramatically cut expenditure to remain viable in an educational system that should be finely tuned and focused on quality education (Reynolds, 2014b). Teacher/professor evaluations must be objective and provide *all* (not just some) teachers/professors with regular feedback that will assist them to grow professionally, no matter how long they have been in the classroom or how close the teacher/professor is to retirement—no one is too old to learn, especially when it comes to the education of young people. In addition, the results of the evaluations should give schools and universities the necessary information to build the strongest possible teaching faculty while weeding out the poor-quality teachers and professors (Stensaker, 2013). In addition, the administrators of schools and universities must be held accountable for assuring the professional development of each teacher.

Evaluations cannot be (but often are) subjective where one of the "good old boys" or "good old girls" is given a good evaluation and the accompanying raise precisely because of the good old boy/girl network. In some instances, it might be the administrator (principal/president/dean) who has reached his/her highest level of incompetence (Peter and Hull, 1969; Lazear, 2000; Pluchino et al., 2010)—often called the *Peter Principle*.

Briefly, the *Peter Principle* is a proposition that states that the members of any organization where promotion is based on achievement, success, and merit will eventually be promoted beyond their level of ability—otherwise referenced by the phrase "employees tend to rise to their level of incompetence" (Peter and Hull, 1969). The principle holds that in a hierarchy, members are promoted so long as they work competently (or for whatever other reasons) until, eventually, they are promoted to a position at which they are no longer competent (their level of incompetence), and there they remain, being unable to earn further promotions. Unfortunately the Peter Principle is incorrect at this point, insofar as some (perhaps, many) teachers and professors continue to receive promotion by managing upward—the concept of a subordinate finding ways to subtly manipulate his/her superiors in order to prevent them from interfering with the subordinate's upward mobility.

In many cases, evaluations are typically perfunctory compliance exercises that rate all teachers/professors as good or exceptional and yield little useful information. This system only serves to frustrate teachers/professors who feel that their good work goes unrecognized and ignores other teachers who would benefit from additional support. The main goal is to avoid the faults and deficiencies of the less than useful "old boy/old girl" evaluation systems or the compliance exercises that only provide false evaluations and assessments. Evaluations should consider multiple *relevant* measures of performance, primarily the impact of the teacher/professor on the scholastic growth of the students.

In addition, the performance evaluation should employ 4–5 rating levels to describe differences in teacher/professor effectiveness and should invoke the concept of regular constructive critical feedback. It serves no purpose for the performance of a teacher/professor to be reflected in the contents of a personnel folder that is two inches thick where 90% of the one-page inserts are complaints from students (and perhaps even complaints from colleagues) and the department head dismissed such complaints as inconsequential. The outcome of any evaluation (and the contents of the personnel folder) must be a major factor in key employment decisions and all teachers/professors should be evaluated at least annually and given clear expectations, which are not subject to misinterpretation.

In fact, at the time of writing, the issue of effective (or ineffective) teachers has already been introduced into the courtroom, leading a California judge in the Los Angeles Superior Court to dismiss the concept of tenure in favor of the rights of the students for a quality education (Medina, 2014a,b). It is expected that the Teachers Union will appeal the decision while other states and cities are likely to follow the legal pathway to negate teacher tenure laws.

A professional evaluation should produce information that can be factored into important decisions about tenure, compensation, development, promotion, and dismissal. In short, the results of any evaluation must be accurate, objective, and easy to interpret. Some administrators may support using evaluation results only to reward good/excellent teachers/professors (such as awarding tenure), and not for decisions such as dismissal—which does not happen (even though it may be warranted) in many schools and universities.

As long as the evaluation process is objective, the results can be used to form the foundation of teacher/professor development, but there *must* be meaningful consequences for consistently poor performance. With better teacher evaluations in place, the leaders of schools and universities can be held accountable for teacher/professor and student performance. In this way, evaluations can play a major role in important employment decisions—teachers and professor exhibiting consistently poor performance as well as the principal/president/dean should be dismissed, which is the way of the outside (nonscholastic) world. Making teachers/professors accountable or responsible—two bad words in many institutions of learning where the focus is an unmitigated authority and a lack of responsibility—for the outcome of their activities is a necessary part of any learning system.

In short, and to make the evaluation process as objective as possible, a teacher/professor evaluation should involve the following criteria: (1) There should be a process in which all teachers should be truthfully evaluated at least annually; (2) there should be clear, rigorous expectations; (3) the evaluation should be based on clear standards of teaching that prioritize student learning; (4) the evaluation should involve multiple measures of performance; (5) the evaluation should employ 4–5 rating levels to describe differences in teacher/professor effectiveness; (6) there should be regular constructive feedback; and (7) there must be significance to the evaluation and the outcome and it should be a major factor in key employment decisions about teachers/professors (TNTP, 2010). Finally, the ultimate success of the evaluation system—no matter how pertinent the design and operation of the system—depends on how well the system is implemented by the educational administrators.

In summary, taking into account the academic history of the student, reasonable attainable goals must be set for the academic progress expected for students during the course of a year. Since the primary professional responsibility of the teacher/professor is to ensure that students learn, measures of student learning should play a predominant role in teacher/professor evaluations. While it may not be the only criterion, it does mean that teachers/professors should be accountable for helping students make measurable progress against the learning standards. Through focused and rigorous observation of classroom practice, examination of student work, and analysis of student performance, it is possible to accurately distinguish effective teaching from ineffective teaching.

## 7.4   CRITERIA FOR SUCCESS

Success requires that there must be a balance between teaching and research with each being equally important to the well-being of the organization and the students (Matthews, 2007). Rapid and extensive improvement of education in science and engineering is unlikely to occur until it becomes clear to scientists and engineers that they have an obligation to become involved in elementary-level and secondary-level science. In fact, it should be proposed that many professors could, with sufficient stimulus and impetus, find the time for such activities.

### 7.4.1   FINANCIAL ASSISTANCE FOR EDUCATION

The concept that every student has the right to attend university has remained endemic in the United States and many other countries where ability to attend university is determined by the ability of the parents of the student to pay the relevant fees and (pass or fail) to keep the student in the university. Come rain, snow, or sunshine, such a student may stay at university until s/he graduates but surely this must raise questions in any profession about the abilities of the professional who has failed courses for several years but, thanks to parental money, is able to remain at university until s/he graduates—say, at age 30 rather than at age 22.

In many countries the phrase "every student has the right to attend university if s/he is good enough" is used, where the term "good enough" is defined or assessed by the scholastic abilities of the would-be university student (Porter, 2014). The term *good enough* is further defined or assessed by the abilities of the student in his/her preuniversity career, which lead to decisions as to whether or not the student will be a scholastic success in a university setting.

On the other hand, there are many university students who are forced (because family finances do not permit support of higher education) to support themselves through university by working in part-time or even full-time jobs. While some observers may consider this a good learning experience, it can detract from the actual learning process, and one must wonder if these same observers were part of the part-time or full-time working force while they were students. Many such students may feel jaded by the time they are awarded the degree or may even leave university because of the stress and strain, which they feel is intolerable and hazardous to health. In such cases, valuable talent may have been lost.

It is therefore preferential that "all preuniversity students be given the chance to attend university, if they are scholastically good enough." A nationwide competitive and realistic system of grants or scholarships for preuniversity students as awards for their preuniversity efforts and to support the students while at university is necessary to develop students that graduate with a high degree of competency in science or engineering. Such a system would ensure that the brightest and most capable young minds have the resources they need to pursue a technologically oriented education. And such a system must be based on awarding degrees in an objective manner so that future employment is based on the ability of the student(s) to obtain a meaningful degree and not on where the student went and for how long (Kamenetz, 2013).

Unlike elementary and secondary schooling, higher education in science and engineering is subject to direct monetary and (especially) federal influence. Moreover high-school students, particularly from low-income backgrounds, have a very poor understanding of the variants in university tuition and opportunities for financial aid, as well as the admissions and financial-aid application processes (Ikenberry and Hartle, 1998; ACSFA, 2001, 2005; Horn, et al., 2003; Kane and Avery, 2004; King, 2004) and this is particularly hurtful if the student is scholastically qualified to enter university. Federal funding is particularly important at the university baccalaureate level and the graduate level, where federal fellowships and other forms of assistance are awarded to support specific graduate students in specific fields of study, and where the majority of academic research is supported by federal funds. Typically, graduate students depend on different sources of support in different phases of graduate work—perhaps as teaching assistants (TAs) in the first two years and then as research assistants while doing dissertation research. By the time students receive the doctorate, nearly two-thirds have been research assistants and half teaching assistants. The students reporting this information are not always sure of the ultimate source of their research-assistant funds, and the reported data do not distinguish between federal and institutional research assistants (National Academies, 2011).

There are guidelines—sometimes clear, often unclear—for distributing the various types of federal support. In fact, the number of doctorates produced often is a reflection of the availability of research funds rather than the employment demand for young men and women holding doctorates. There are several drawbacks to this dependence on research grants, such as the continued pressure to produce new research results; this extends from the professor (the grant holder) to graduate students, who easily gain the impression that hard, goal-oriented work on a specific project is the most valuable aspect of graduate education.

As already noted, doctoral students can become so involved in the work of the faculty investigators under whose grants they conduct their dissertation research that little time is left for independent exploration or other educational activities. Even the best-intentioned professors might lack the time to impart a broad appreciation of their discipline or to encourage their doctoral students to investigate the discipline thoroughly or plan their careers. Efforts need to continue to be made to make this experience as profitable and broadening as possible so that graduate scientists and engineers are prepared for all type of careers.

## 7.4.2 A MULTIDISCIPLINARY APPROACH

The high-school science curriculum was first developed toward the end of the nineteenth century in Europe and North America and the curriculum was organized around the administrative units of physics, chemistry, biology, and geology within the various universities. However, over the past 100 years, science and engineering have changed considerably—they have emerged from the dark ages of university thought of the nineteenth century with little focus on the needs of industry and moved into enlightened and integrated relationships with industry, the military, and other technological groups (i.e., in the real world). At the same time, the disciplines of science and engineering have evolved into a multitude of subdisciplines which are further integrated with combinations of older and newer disciplines.

Just as science and engineering should be multidisciplinary areas of study, high-school science and engineering (i.e., mathematics) curricula have not kept pace and the compartmentalized (so-called unchangeable) disciplines of pure physics, chemistry, and biology, along with botany and zoology, have remained virtually untouched. Thus it is not surprising that multidisciplinary approaches to problem solving have been held back, unless specifically ordered as teamwork by a higher authority. As a result, there has been the call for a multidisciplinary approach as well as reforms to the education system for science and engineering education in high schools (Solomon and Aikenhead, 1995; Aikenhead, 1997).

## 7.4.3 ACHIEVEMENTS AND RECOGNITION

The scientific and engineering fields are composed of educated and relatively young professionals who have the ability to apply themselves to the problems at hand, whether theory studies or experimentation. To the scientist and engineer, the outcome of this work that offers some form of gratification is (1) the completion of a project, and (2) publication of the data in a journal or company-related publication medium for distribution to one's peers. The last gives the scientist or engineer intercompany and intracompany recognition for his/her work.

Moreover, in regard to the publication of theoretical concepts or experimental data, scientific and engineering professionals who are biased toward the theoretical aspects of their respective disciplines tend to produce abstract concepts in which the intellectual contribution is expressed in the form of theories, with or without tangible proof. As a result, publication on the proceedings of a conference may be the predominant outlet for such efforts after which publication in a *reputable* journal may be possible, but only with considerable efforts; or for various reasons, publication in a reputable journal may not be possible at all. For the nonacademic scientist and engineer, there is the medium of publication of the material as a *company report*. This can be a worthwhile method for circulating one's work throughout the company. But the importance of the work to the young scientist and engineer can, again, be diminished if the names of a supervisor and any other persons higher up the food chain are included as coauthors.

Presentation of the work at a conference followed by publication of the data in the proceedings from a conference often results in a shorter time to data release and to print, but in terms of speaking and writing, the presentation must be logical and understandable (Walters and Walters, 2011; Laplante, 2012). This follows from the opportunity to describe competed or partly completed work before peer scientists and/or engineers and to receive a more complete review than the type of review that is typical for a journal. At a conference, the audience asks general and specific questions of the presenter that often guide the presenter into further work or to diverge into a new line of investigation. Overall, this will help the presenting scientist/engineer to finalize the document for publication in the proceedings (where the proceedings are published postconference).

However, publication in a scientific or engineering journal is not always the zenith of achievement. There is always the concern that journal reviewers do not really pay attention to the salient points of the potential publication but pay more attention to the big name (the prominent scientist or prominent engineer) among the authors. Some reviewers miss all of the salient facts as well as the big-name author and they merely look for errors in style and grammar. Several scientists and engineers may have experienced all of the above behavior on the part of reviewers. In many cases in academia, statements are made that publication in the proceedings of a prestige conference is inferior to publication in a prestige journal, without the professors realizing or being willing to admit that in relation to data presentation and publication, the technical content of many conferences is superior to the technical content of some so-called established journals. The journal should actually be at the forefront of reliable and honest knowledge (monitored by the editor and the peer reviewers) in terms of integrity in academic research (Shaw and Despota, 2013).

The scientist or engineer who wishes to publish his/her work may be unable to because of (1) a company policy related to proprietary material—a justified reason— or (2) an arbitrary decision by a supervisor or a member of the company review committee—an unjustified reason. On the other hand, in academia, the young professional enters a department at the assistant professor grade in which there is little (or no) choice in terms of choosing teaching assignments and has administrative work thrust upon his/her shoulders, while the older tenured members of staff have the shameless boldness to refuse such work without fear of reprisal under the umbrella of being tenured. In short, there should be a complete overhaul of the tenure system in universities (and schools) with the real option that it should actually be discontinued. The original idea behind tenure has been corrupted to such an extent that it is of no benefit to the education system.

In addition, the young assistant professor also has to acquire external funding for his/her area of research and may even have to pass his/her reports/papers through a review committee prior to publication. This review committee will be made up of senior members of staff who, for many reasons that are often difficult to follow, can give the young professor a promising performance report or a report that is somewhat less than glowing. It is at this time, if the latter is the case, that the young professor can feel that s/he is suffering rejection by his/her colleagues. In an industrial setting the same can happen and the educated young professional scientist and engineer wonders if s/he is merely a pair of hands (for an overbearing supervisor or an overbearing department head or jealous colleagues, who will nevertheless not stop short of claiming some form of contribution if the research work is successful) and is not supposed to be given credit for the ability to think and solve a problem. Performance suffers and, with repeated negativism toward publication, the young professional starts to lose interest in the organization.

Lack of recognition for hard and intelligent work is a killer for young scientists or engineers, and getting the best out of any such scientists and engineers becomes an impossible dream or a self-fulfilling prophecy, due to the behavior of the malicious miscreants.

## 7.5  OVERHAULING THE CONCEPT OF TENURE

Tenure is a major hurdle to improving the education system in schools and universities.

## 7.6  A NATIONAL POLICY

In an attempt to introduce national standards for education, the Common Core Standards have been introduced (http://www.corestandards.org/read-the-standards/). Successful implementation of these standards requires that parents, teachers, and policymakers have the facts about development, intent, content, and implementation of the standards. The standards are designed to build upon the most advanced current thinking about preparing all students for success in college, career, and life. This will result in moving even the best state standards to the next level. In fact, since this work began, there has been an explicit agreement that no state would lower its standards. The standards will serve to produce students that are university-ready.

In addition, the major educational-funding participants—universities, government, industry, and any other funding organizations—must agree that change is inevitable and that the move toward globalization has begun, (Burland et al., 2004; Chubin et al., 2005). In fact, globalization has been in place for decades and is reflected in the migration (often referred to as the *brain drain* by those countries from which the scientists and engineers migrated) of many European scientists and engineers to Western countries, especially the United States.

In order for the United States to regain the technological edge that it had in the 1970s, it is necessary to start at the time when potential scientists and engineers commence education in the various preuniversity school systems. In addition, there must also be agreement on what is required for teachers to be qualified (accredited) and what it takes to prepare teachers for the tasks at hand (Chapter 1). Answers—and opinions about the answers—are not in short supply and decisions about teacher preparation and accreditation are made on a variety of bases. For example, within the United States, teachers are prepared in numerous (>1300) large and small, public and private colleges and universities, as well as through alternative programs offered by school districts and states. Admirable as this may seem (in terms of the numbers), the lack of a meaningful national policy results in differences in programs and teacher preparation (Wilson et al., 2001).

Finally, although the population of school-age children in the United States is becoming increasingly diverse, the pool of potential teachers is not equally diverse. A qualified teaching force is an unquestionable necessity and the diversity of student backgrounds underscores the need to prepare teachers to work with students different from themselves. Issues related to course subject matter, teacher preparation, teacher training, policy influences (by well-meaning or interfering segments of governments), and alternative certification need to be addressed and answered—not just in words but in deeds. The results will provide directions by which to improve teacher qualifications on a national basis (Wilson et al., 2001) leading to progress in the early education of scientist and engineers. The proof of the existence of qualified teachers in schools will be the result of the introduction (and the results) and application of objective evaluation systems.

However, the required teaching qualifications and means of accreditation of the faculty must also be introduced. Qualified (and accredited) faculty members are a necessity for teaching science and engineering and the faculty must be evaluated by the appropriate standards. As with schools, the proof of the existence of qualified teaching faculty in universities will also be evident from the results of the introduction and application of objective evaluation systems.

## REFERENCES

ACSFA (Advisory Committee on Student Financial Assistance). 2001. *Access Denied: Restoring the Nation's Commitment to Equal Educational Opportunity.* Advisory Committee on Student Financial Assistance, Department of Education, Washington, DC.

ACSFA (Advisory Committee on Student Financial Assistance). 2005. The student aid gauntlet: Making access to college simple and certain. Advisory Committee on Student Financial Assistance. Final Report of the Special Study of Simplification of Need Analysis and Application for Title IV Aid. Department of Education, Washington, DC.

Aikenhead, G.D.S. 1997. Integrating the Scientific Disciplines in Science Education. Keynote Address. Gesellschaft fur der Chemie und Physik, Universitat Potsdam, Potsdam, Germany, September 22.

Anderson, E. and Kim, D. 2006. *Increasing the Success of Minority Students in Science and Technology.* American Council on Higher Education, Washington, DC.

ASEE (American Society of Engineering Education). 2009. *ASEE Connections.* American Society of Engineering Education, Washington, DC.

Balf, T. 2014. The SAT is not fair. *New York Times Magazine.* Sunday, March 9, pp. 26–51.

Basken, P. 2011. Scientists fault universities as favoring research over teaching. *Chronicle of Higher Education,* January 13. http://chronicle.com/article/Scientists-Fault-Universities/125944/, accessed January 15, 2014.

Burland, D.M., Doyle, M.P., Rogers, M.E., and Masciangioli, T.M. (eds). 2004. *Preparing Chemists and Chemical Engineers for a Globally Oriented Workforce: A Workshop Report to theChemical Sciences Roundtable.* National Research Council, Washington, DC.

Cheung, J. 2014. The fading honor code: If students fail to report cheaters, can a culture of integrity prevail. *Education Life, New York Times.* April 13, pp. 25–26.

Chubin, D.E., May, G.S., and Babco, E.L. 2005. Diversifying the engineering workforce. *Journal of Engineering Education,* 34: 73–86.

Cockburn, C. 1985. The material of male power. In: M. Donald and J. Wajcman (eds), *The Social Shaping of Technology,* pp. 125–146. Open University Press, Milton Keynes, United Kingdom.

Conefrey, T. 2001. Sexual discrimination and women's retention rates in science and engineering programs. *Feminist Teacher,* 13(3): 170–192.

Cuny, J. and Aspray, W. 2000. Recruitment and retention of women graduate students in computer science and engineering. Report of a Workshop. June 20–21, 2000. Organized By: The Computing Research Association's Committee on the Status of Women in Computing Research. Supported in part by the National Science Foundation Grant No. EIA-9812240 awarded to the Computing Research Association for the study on the supply of information technology workers.

Etzkowitz, H., Kemelgor, C., Neuschatz, M., and Uzzi, B. 1994. Barriers to women in academic science and engineering. In: W. Pearson Jr. and I. Fechter (eds), *Who Will Do Science? Educating the Next Generation*. Johns Hopkins University Press, Baltimore, MD.

Faulkner, W. 2000. Hierarchies and gender in engineering. *Social Studies of Science*, 30(5): 759–792.

Felder, R.M., Woods, D.R., Stice, J.E., and Rugarcia, A. 2000. The future of engineering education II. Teaching methods that work. *Chemical Engineering Education*, 34(1): 26–39.

Fiegener, M.K. and Proudfoot, S.L. 2013. Baccalaureate origins of U.S.-trained S&E doctorate recipients. Survey of earned doctorates. National Science Foundation, National Center for Science and Engineering Statistics, Arlington, VA.

Frehill, L.M. 1997. Education and occupational sex segregation: The decision to major in engineering. *Sociological Quarterly*, 38(2): 225–249.

Getz, M. and Siegfried, J.J. 2010. What does intercollegiate athletics do to or for colleges and universities? Paper No. 1005. Working Papers, Vanderbilt University Department of Economics, Vanderbilt University, Nashville, TN.

Gill, J., Mills, J., and Ayre, M. 2010. *Gender Inclusive Engineering Education*. Routledge, Taylor & Francis Group, Florence, KY.

Hacker, S. 1989. *Pleasure, Power, and Technology: Some Tales of Gender, Engineering and the Cooperative Workplace*. Unwin Hyman, Boston, MA.

Hall, L.E. 2007. *Who's Afraid of Marie Curie? The Challenges Facing Women in Science and Technology*. Seal Press, Emeryville, CA.

Hathaway, R.S., Sharp, S., and Davis, C. 2001. Programmatic efforts affect retention of women in science and engineering. *Journal of Women and Minorities in Science and Engineering*, 7: 107–124.

Hersh, M. 2000. The changing position of women in engineering worldwide. *IEEE Transactions of Engineering Management*, 47(3): 345–359.

Horn, L.J., Chen, X., and Chapman, C. 2003. Getting ready to pay for college: What students and their parents know about the cost of college tuition and what they are doing to find out. National Center for Education Statistics Report No. 2003030. National Center for Education Statistics, Washington, DC.

Hunt, J. 2010. Why do women leave science and engineering? NBER Working Paper 15853. National Bureau of Economic Research, Cambridge, MA.

IChemE (Institution of Chemical Engineers). 2013. http://www.icheme.org/media_centre/news/2013/model-predicts-risk-of-women-engineers-leaving-profession.aspx#.UffAnY3VA1M, accessed July 30, 2013.

Ikenberry, S.O. and Hartle, T.W. 1998. Too little knowledge is a dangerous thing: What the public thinks about paying for college. Report. American Council on Education, Washington, DC.

Jacobs, J.A. and Winslow, S.E. 2004. The academic life course, time pressures and gender inequality. *Community, Work and Family*, 7(2): 143–161.

Kamenetz, A.M. 2013. Are you competent? Prove it. *Education Life, New York Times*. November 3, pp. 26–27.

Kane, T.J. and Avery, C. 2004. Student perceptions of college opportunities: The Boston COACH program. In: C. Hoxby (ed.), *College Decisions: The New Economics of Choosing, Attending and Completing College*. University of Chicago Press, Chicago, IL.

King, C.J. 2011. Restructuring engineering education: Why, how and when? Research and occasional paper series: CSHE.12.11. Center for Studies in Higher Education (CSHE), University Of California, Berkeley, CA.

King, J.E. 2002. *Crucial Choices: How Students' Financial Decisions Affect Their Academic Success*. American Council on Education, Washington, DC.

King, J.E. 2004. Missed opportunities: Students who do not apply for financial aid. Report— Issue Brief. American Council on Education, Washington, DC.

Kvande, E. 1999. In the belly of the beast—Constructing femininities in engineering organizations. *The European Journal of Women's Studies*, 6(3): 305–328.

Kvande, E. and Rasmussen, B. 1994. Men in male-dominated organizations and their encounter with women intruders. *Scandinavian Journal of Management*, 10(2): 164–175.

Lackland, A.C. and De Lisi, R. 2001. Students' choices of college majors that are gender traditional and nontraditional. *Journal of College Student Development*, 42(1): 39–48.

Laplante, P.A. 2012. *Technical Writing: A Practical Guide for Scientists and Engineers*. CRC Press, Taylor & Francis Group, Boca Raton, FL.

Lazear, E.P. 2000. The Peter Principle: Promotions and declining productivity. Revision 10/12/00. Hoover Institution and Graduate School of Business, Stanford University, Stanford, CA. http://www-siepr.stanford.edu/Papers/pdf/00-04.pdf, accessed July 22, 2013.

Lewin, T. 2013. Student debt grows faster at universities with highest-paid leaders. *Education Life, New York Times*. August 4, pp. 16–17.

Lewin, T. 2014. Sneak preview: What the new SAT and digital ACT might look like. *New York Times*. May 19, p. A11.

Lewis, S. 1993. Including gender in higher education science and engineering courses. In *Proceedings of Gender Science and Technology (GASAT) 7th International Conference: Transforming Science and Technology: Our Future Depends on it*, pp. 662–669. University of Waterloo, Ontario, Canada.

Marx, G. 1990. Reflections on academic success and failure: Making it, forsaking it, reshaping it. In: B. Berger (ed.), *Authors of Their Own Lives*, pp. 260–284. University of California Press, Berkeley, CA.

Matthews, C.M. 2007. Science, engineering, and mathematics education: Status and issues. Report No. 98-871 STM. Update April 23, 2007. CRS Report for Congress, Congressional Research Service, Washington, DC.

McBay, S.M. 1989. Improving education for minorities. *Issues in Science and Technology*, 5(4): 41–47. http://educationinaminute.com/bogota/Education%20Reports/McBay%20 -%20Article%20(Improving%20educ%20for%20Minorities).pdf.

McIlwee, S. and Robinson, G. 1992. *Women in Engineering: Gender, Power, and Workplace Culture*. State University of New York Press, Albany, NY.

Medina, J. 2014a. Fight over effective teachers shifts to courtroom. *New York Times*. February 1, pp. A1,A13.

Medina, J. 2014b. Judge rejects teacher tenure for California: Rules system deprives students of rights. *New York Times*. February 1, pp. A1,A15.

Morgan, L.A. 2000. Is engineering hostile to women? An analysis of data from the 1993 National Survey of College Graduates. *American Sociological Review*, 65(2): 316–321.

NAP (National Academies Press). 1994. *Women Scientists and Engineers Employed in Industry: Why So Few? Committee on Women in Science and Engineering*. National Research Council, National Academies Press, Washington, DC.

NAP (National Academies Press). 1995. *Reshaping the Graduate Education of Scientists and Engineers*. National Academies Press, National Academy of Sciences, National Academy of Engineering, Institute of Medicine, Washington, DC.

NAS (National Academy of Sciences). 1997a. *Adviser, Teacher, Role Model, Friend: On Being a Mentor to Students in Science and Engineering.* National Academy of Sciences, National Academy of Engineering, Institute of Medicine, National Academy Press, Washington, DC.

NAS (National Academy of Sciences). 1997b. *Transforming Undergraduate Education in Science, Mathematics, Engineering, and Technology.* National Academy of Sciences, National Academy of Engineering, Institute of Medicine, National Academy Press, Washington, DC.

NAS (National Academy of Sciences). 2000. *Enhancing the Postdoctoral Experience for Scientists and Engineers: A Guide for Postdoctoral Scholars, Advisers, Institutions, Funding Organizations, and Disciplinary Societies.* National Academy of Sciences, National Academy of Engineering, Institute of Medicine, National Academy Press, Washington, DC.

NAS (National Academy of Sciences). 2005. *Educating the Engineer of 2020: Adapting Engineering Education to the New Century.* National Academy of Sciences, National Academy of Engineering, Institute of Medicine, National Academy Press, Washington, DC.

National Academies. 2011. A Data-Based Assessment of Research-Doctorate Programs in the United States. National Academies Press, Washington, DC.

NRC (National Research Council). 1997. *Science Teaching Reconsidered: A Handbook.* National Research Council, National Academies Press, Washington, DC.

NRC (National Research Council). 2003. Evaluating and Improving Undergraduate Teaching in Science, Technology, Engineering, and Mathematics. Committee on Recognizing, Evaluating, Rewarding, and Developing Excellence in Teaching of Undergraduate Science, Mathematics, Engineering, and Technology, M.A. Fox and N. Hackerman (eds), Center for Education, Division of Behavioral and Social Sciences and Education. The National Academies Press, Washington, DC.

NSF (National Science Foundation). 1996. Shaping the future: New expectations for undergraduate education in science, mathematics, engineering, and technology. Report No. NSF 96-139. National Science Foundation, Washington, DC.

NSF (National Science Foundation). 2012. Doctorate recipients from U.S. Universities: 2011. Report No. NSF 13-301. National Science Foundation, National Center for Science and Engineering Statistics (NCSES), Arlington, VA, December. http://www.nsf.gov/statis tics/doctorates/, accessed August 7, 2013.

OECD (Organization for Economic Cooperation and Development). 2012. *Education at a Glance 2012: Highlights.* OECD Publishing, Organization for Economic Cooperation and Development, Washington, DC. http://www.oecd-ilibrary.org/education/education- at-a-glance-2012_eag_highlights-2012-en, accessed July 22, 2013; http://www.oecd. org/education/skills-beyond-school/48631419.pdf, accessed July 22, 2013.

Osborne, J. 2007. Science education for the twenty first century. *Eurasia Journal of Mathematics, Science and Technology Education*, 3(3): 173–184.

Peter, L.J. and Hull, R.R. 1969. *The Peter Principle: Why Things Always Go Wrong.* William Morrow and Company, New York.

Pluchino, A., Rapisarda, A., and Garofalo, C. 2010. The Peter Principle revisited: A computational study. *Physica A: Statistical Mechanics and its Applications*, 389(3): 467–472.

Porter, E. 2014. For schools, long road to a level playing field. *Business Day, New York Times.* May 21, pp. B1,B6.

Preston, A.E. 1994. Why have all the women gone? A study of exit of women from the science and engineering professions. *American Economic Review*, 84: 1446–1462.

Reynolds, G.H. 2014a. Higher Ed Sports lower standards. *USA Today—Hawaii Edition.* January 15, p. 9A.

Reynolds, G.H. 2014b. Degrees of value—Making college pay off. *The Wall Street Journal.* Saturday/Sunday Edition, January 4–5, p. C1.

Shaw, M.M. and Despota, K. 2013. Journals: At the front line of integrity in academic research. In: *Global Corruption Report: Education* (Chapter 3.20). Routledge, Taylor & Francis Group, Abingdon, Oxford.

Shaywitz, S. and Hahm, J. (eds). 2004. *Achieving XXcellence in Science: Role of Professional Societies in Advancing Women in Science—Proceedings of a Workshop, AXXS 2002 Committee on Women in Science and Engineering.* National Research Council, Washington, DC.

Singh, R., Fouad, N.A., Fitzpatrick, M.E., Liu, J.P., Cappaert, K.J., and Figuereido, C. 2013. Stemming the tide: Predicting women engineers' intentions to leave. *Journal of Vocational Behavior*, 83(3): 281–294.

Solomon, J. and Aikenhead, G.S. 1995. *STS Education: International Perspectives on Reform.* Vintage Books, New York.

Sonnert, G. 1990. Signs: The elusive concepts. Journal of Women in Culture and Society. Proceedings. Careers of Women and Men Postdoctoral Fellows in the Sciences. American Sociological Association Meetings, August. Role Models, Mentors and Sponsors, vol. 6, no. 41, pp. 692–712.

Speight, J.G., and Foote, R. 2011. *Ethics in Science and Engineering.* Scrivener, Salem, MA.

Stensaker, B. 2013. Ensuring quality in quality assurance. In: *Global Corruption Report: Education* (Chapter 3.5). Routledge, Taylor & Francis Group, Abingdon, Oxford.

Sung, N.S., Gordon, J.I., Rose, G.D., Getzoff, E.D., Kron, S.J., Mumford, D., Onuchic, J.N., Scherer, N.F., Sumners, D.L., and Kopell, N.J. 2003. Educating future scientists. *Science*, 301: 1485. www.sciencemag.org/cgi/content/full/301/5639/1485, accessed June 1, 2014.

Tapia, R.A. 2009. Minority students and research universities: How to overcome the mismatch. *The Chronicle of Higher Education*, 55(29): A72. http://chronicle.com/weekly/v55/i29/29a07201.htm, accessed July 29, 2013.

TNTP (The New Teacher Project). 2010. *Teacher Evaluation 2.0. The New Teacher Project.* TNTP, Brooklyn, NY. http://tntp.org/assets/documents/Teacher-Evaluation-Oct10F.pdf?files/Teacher-Evaluation-Oct10F.pdf, accessed August 12, 2013.

Tuhus-Dubrow, R. 2013. The repurposed PhD: Finding life after academia and not feeling bad about it. *Education Life, New York Times*. November 3, pp. 32–33.

Walters, D.E. and Walters, G.C. 2011. *Scientists Must Speak*, 2nd edn. CRC Press, Taylor & Francis Group, Boca Raton, FL.

Washburn, M.H. and Miller, S.G. 2005. Retaining undergraduate women in science, engineering, and technology: A survey of a student organization. *Journal of College Student Retention*, 6(2): 155–168.

Weingarten, R. 2013. Drowning in debt to get a degree. *Sunday Review, New York Times*. June 16, p. 2.

Wilson, S.M., Floden, R.E., and Ferrini-Mundy, J. 2001. Teacher preparation research: Current knowledge, gaps, and recommendations. Document R-01-03. Educational Research and Development Centers Program, PR/Award Number R308B970003, as administered by the National Institute on Educational Governance, Finance, Policymaking and Management, Office of Educational Research and Improvement (OERI), U.S. Department of Education, Washington, DC.

Winerip, M. 2013. Children of the state—Students on their own. *Education Life, New York Times*. November 3, pp. 20–23.

Winerip, M. 2014. Stepping up to stop sexual assault. *Education Life, New York Times*. February 9, pp. 14–17.

# 8 Glossary

**academic performance index (API):** A statewide ranking of schools based on student test scores from the CAT/6, CST, and high-school exit exam; it ranges from 200 to 1000. Most schools have an API, a state ranking (by elementary, middle, or high school), a ranking in comparison to 100 similar schools, and growth targets for the following year.

**academic year:** The university year, which usually runs from September to June and is divided into two semesters (or three terms) of teaching, with vacations at Christmas and Easter.

**accommodations:** Changes in the way tests are designed or administered to respond to the special needs of students with disabilities and English learners (EL).

**accountability:** The notion that people (i.e., students or teachers) or an organization (i.e., a school, school district, college of education, university or any educational establishment) should be held responsible for improving student achievement and should be rewarded or sanctioned for their success or lack of success in doing so.

**achievement gap:** The gap in performance criteria that occurs when one group of students outperforms another, where the difference in average scores for the two groups is statistically significant (i.e., larger than the margin of error).

**achievement test:** A test to measure a student's knowledge and skills.

**American college testing (ACT):** A set of admissions tests for entry into a college, university, or an establishment of higher education. Most establishments of higher education accept either the SAT or the ACT for admissions purposes.

**adequate yearly progress (AYP):** An individual state's measure of yearly progress toward achieving state academic standards. Adequate yearly progress is the minimum level of improvement that states, school districts, and schools must achieve each year.

**adoption:** Refers to the chosen curriculum of a particular school.

**advanced:** See *Proficiency.*

**advanced placement (AP):** A series of voluntary exams based on university/college-level courses taken in high school. High-school students who do well on one or more of these exams have the opportunity to earn credit, advanced placement, or both for higher education.

**advancement via individual determination (AVID):** A 4-year elective college/university preparatory class designed to motivate students to attend college.

**alignment:** The degree to which assessments, curriculum, instruction, textbooks and other instructional materials; teacher preparation and professional development; and systems of accountability all reflect and reinforce the educational program's objectives and standards.

**alternative assessments:** Ways other than standardized tests to get information about what students know and where they need help, such as oral reports, projects, performances, experiments, and class participation.

**alternative schools accountability model (ASAM):** An alternative way of measuring student performance in schools with mostly high-risk students—such as continuation schools or some county office of education schools—and schools with fewer than 11 valid test scores.

**alumni:** Former students.

**annual measurable objective (AMO):** The annual target for the percentage of students whose test scores must be proficient or above in English/language, arts, and mathematics.

**assessment:** Teacher-made tests, standardized tests, or tests from textbook companies that are used to evaluate student performance.

**assistant professor (rank):** A faculty member at a university who is at the entry-level teaching rank; typically a nontenured position.

**associate degree:** A degree granted for the successful completion of a subbaccalaureate program of studies, usually requiring at least 2 years (or equivalent) of full-time college-level study; this includes degrees granted in a cooperative or work-study program.

**associate professor (rank):** A faculty member at a university who is at the mid-level teaching rank; typically a tenured position.

**at-risk student:** Students may be labeled at risk if they are not succeeding in school based on information gathered from test scores, attendance, or discipline problems.

**average class size:** The number of students in classes divided by the number of classes. Because some teachers, such as reading specialists, have assignments outside the regular classroom, the average class size is usually larger than the pupil–teacher ratio.

**average daily attendance (ADA):** The total number of days of student attendance divided by the total number of days in the regular school year. A student attending every school day would equal one ADA. Generally, ADA is lower than enrollment due to such factors as transiency, dropouts, and illness. A school district's revenue limit income is based on its ADA.

**baccalaureate:** An academic degree conferred by a college or university upon those who have completed the undergraduate curriculum as a registered student; also called a *bachelor's degree.*

**basic:** See also *Proficiency.*

**basic aid:** The minimum general-purpose aid that is guaranteed by the state's constitution for each school district in California. A basic-aid district is one in which local property taxes equal or exceed the district's revenue limit. These districts may keep the money from local property taxes and still receive constitutionally guaranteed state funding.

**benchmarks:** A detailed description of a specific level of student achievement expected of students at particular ages, grades, or developmental levels; academic goals set for each grade level.

**bilingual education:** An in-school program for students whose first language is not English or who have limited English skills. Bilingual education provides English language development plus subject-area instruction in the student's native language. The goal is for the child to gain knowledge and be literate in two languages.

**block scheduling:** Instead of traditional 40–50-min periods, block scheduling allows for periods of an hour or more so that teachers can accomplish more during a class session. It also allows for teamwork across subject areas in some schools. For example, a math and science teacher may teach a physics lesson that includes both math and physics concepts.

**bond measure:** A method of borrowing used by school districts to pay for construction or renovation projects. A bond measure requires a 55% majority to pass. The principal and interest are repaid by local property owners through an increase in property taxes. (See also *parcel tax*.)

**categorical aid:** Funds from the state or federal government granted to qualifying schools or districts for specific children with special needs, certain programs such as class-size reduction, or special purposes such as transportation. In general, schools or districts must spend the money for the specific purpose. All districts receive categorical aid in varying amounts. This aid is in addition to the funding schools receive for their general education program.

**certified/credentialed employees:** School employees who are required by the state to hold teaching credentials, including full-time, part-time, substitute, or temporary teachers and most administrators. A teacher who has not yet acquired a credential but has an emergency permit or a waiver to teach in the classroom is included in the count.

**certificate/credential:** A state-issued license certifying that the teacher has completed the necessary basic training courses and passed the teacher exam.

**charter schools:** Publicly funded schools that are exempt from many state laws and regulations for school districts. They are run by groups of teachers, parents, and/or foundations.

**class-size reduction:** A state-funded program for kindergarten through third grade classes to ensure that there are no more than 20 students per teacher. A separate program supports some smaller classes for core subjects in ninth grade.

**classified employees:** School employees who are not required to hold teaching credentials, such as bus drivers, secretaries, custodians, instructional aides, and some management personnel.

**closed campus:** This usually indicates that the school has one point of entry and a sign-in procedure as safety measures. It also refers to a high school that does not allow students to leave the campus for lunch or does not allow students to come and go without permission during the school day.

**cluster:** To place small groups of students together for instruction, especially *GATE* students.

**college:** An institute of higher education; a term sometimes used erroneously to describe a university, which may in reality be a collection of colleges.

**community college:** A 2-year college, also referred to as a junior college. Anyone who is 18 years old or holds a high-school diploma (or equivalent) is eligible to attend a community college.

**community-based learning:** Students, faculty, administrators, and community members working together to create new learning opportunities within local communities but generally outside traditional learning institutions.

**comparable growth:** Subgroups of students in a school must improve their scores on standardized tests. They are expected to achieve 80% of the predominant student group's target, which is known as comparable growth.

**conflict management:** A strategy that schools use to prevent and address behavior problems by using peer counselors, mediators, or a school curriculum or program. It usually includes a set of expectations for behavior.

**consolidated application (Con App):** The application districts can use to apply for more than 20 state and federal categorical programs, including the federal Title I program and the state School Improvement Program (SIP). Most, if not all, districts use the Con App to secure funding from at least some of the programs on the application.

**content standards:** Standards that describe what students should know and be able to do in core academic subjects at each grade level.

**cooperative learning:** A teaching method in which students of differing abilities work together on an assignment. Each student has a specific responsibility within the group. Students complete assignments together and receive a common grade.

**core academics:** The required subjects in middle and high schools—usually English (literature), history (social studies), math, and science, each of which may be housed in a separate university college.

**criterion-referenced test:** A test that measures how well a student has learned a specific body of knowledge and skills. The goal is typically to have every student attain a passing mark, rather than to compare students to each other. (See *norm-referenced assessment*.)

**crosscultural language and development (CLAD):** A test that teachers must pass to gain credentials that qualify them to teach English to English learners. The BCLAD is a CLAD for bilingual teachers.

**curriculum:** The courses of study offered by a school or district. California has developed a set of standards that are intended to guide curriculum and instruction. The final decisions about school curriculum are the responsibility of the local school board.

**degree-granting institutions:** Postsecondary institutions that are eligible for Title IV federal financial-aid programs and grant an associate's or higher degree. For an institution to be eligible to participate in Title IV financial-aid programs it must offer a program of at least 300 clock hours in length, have accreditation recognized by the U.S. Department of Education, have been in business for at least 2 years, and have signed a participation agreement with the Department of Education.

**differentiated instruction:** Also referred to as *individualized instruction* or *customized instruction*. The curriculum offers several different learning experiences within one lesson to meet students' varied needs or learning styles. For example, different teaching methods are offered for students with learning disabilities.

**diploma:** A nondegree offering below the associate degree, which is most often offered in technical and vocational fields of study; the diploma generally leads to employment in an occupational field, such as certificate in bookkeeping and certificate in automotive technology.

**disaggregated data:** The presentation of data broken into segments of the student population instead of the entire enrollment. Typical segments include students who are economically disadvantaged, from racial or ethnic minority groups, have disabilities, or have limited English fluency. Disaggregated data allows parents and teachers to see how each student group is performing in a school.

**Doctor's degree (*doctorate*):** An earned degree that generally carries the title of Doctor. The Doctor of Philosophy degree (PhD), the Doctor of Science degree (DSc), and the Doctor of Engineering Degree (DEng) are the highest academic degrees and require mastery within a field of knowledge and demonstrated ability to perform scholarly research. Other doctor's degrees are awarded for fulfilling specialized requirements in professional fields, such as education (EdD), musical arts (DMA), business administration (DBA), and engineering (DEng or DES). Many doctor's degrees in academic and professional fields require an earned master's degree as a prerequisite. The doctor's degree classification includes most degrees that NCES formerly classified as first-professional degrees. Such degrees are awarded in the fields of dentistry (DDS or DMD), medicine (MD), optometry (OD), osteopathic medicine (DO), pharmacy (PharmD), podiatry (DPM, PodD., or DP), veterinary medicine (DVM), chiropractic (DC or DCM), and law (LLB or JD).

**dropout:** A grade 7–11 student who left school prior to completing the school year and had not returned by Information Day (a day in October when students throughout the state are counted and enrollment is determined). This does not include students who receive a General Education Development (GED) or California High School Proficiency Examination (CHSPE) certificate, transfer to another high school or to a college, move out of the United States, are suspended or sick that day, or enrolled late.

**elementary school:** A school classified as elementary by state and local practice and composed of any span of grades not above Grade 8.

**emergency permit:** In California, a 1-year permit issued to people entering the teaching profession who have not completed some of the legal requirements for a credential. Generally the intent is that the person will enroll in and complete an approved teacher-preparation program.

**English as a second language classes:** Support programs for students whose native language is not English.

**English Language Advisory Committee (ELAC):** Variations include *English Language Advisory Council* and *English Language Learner Advisory Committee/Council*. The group consists of parents and school staff who work together to address the academic needs of students still learning English.

**English learner:** A student who is not proficient enough in the English language to succeed in the school's regular instructional programs and who qualifies for extra help.

**enrichment:** Additional courses outside those required for graduation.

**family math (*family mathematics*):** A program that teaches families how to enjoy doing math together. Parents and children attend workshops or use the family math book to learn how to use everyday materials to do fun math activities.

**fluent English proficient (FEP):** A designation that means that a student is no longer considered as part of the school's English learner population. It refers to students who have learned English.

**formative assessment:** Any form of assessment used by an educator to evaluate students' knowledge and understanding of particular content and then to adjust instructional practices accordingly toward improving student achievement in that area.

**free/reduced-price meals:** A federal program that provides food for students from low-income families.

**general fund:** Accounting term used by the state and school districts to differentiate general revenues and expenditures from funds for specific uses, such as a cafeteria fund.

**gifted and talented education (GATE):** A program that offers supplemental, differentiated, and challenging curriculum and instruction for students identified as being intellectually gifted or talented.

**governor's performance awards:** A competitive program that grants awards to public schools in California that meet or exceed the Academic Performance Index performance-growth target each year.

**graduate enrollment:** The number of students who are working toward a master's or doctor's degree. These enrollment data measure those students who are registered at a particular time during the fall. At some institutions, graduate enrollment also includes students who are in postbaccalaureate classes but not in degree programs. In most tables, graduate enrollment includes all students in regular graduate programs and all students in postbaccalaureate classes but not in degree programs (unclassified postbaccalaureate students).

**highly qualified teacher:** A teacher who has obtained full state-teacher certification or has passed the state-teacher licensing examination and holds a license to teach in the state; holds a minimum of a bachelor's degree; and has demonstrated subject-area competence in each of the academic subjects in which the teacher teaches.

**high priority schools grant program (HPSGP):** A program created to provide funds for schools in the lower half of the state rankings (Deciles 1–5) based on the API. It focuses on schools with APIs that fall in the bottom 10% of all schools and replaces the *II/USP*. Schools volunteer to be in this program.

**II/USP (immediate intervention/underperforming schools program):** The Immediate Intervention/Underperforming Schools Program was designed to encourage a school-wide improvement program in schools with very low test scores and to provide assistance and intervention. Schools in the lowest five deciles of API scores were eligible if they did not meet their API targets. It was replaced in 2002 with HPSGP, a similar program.

**immersion education:** A program that teaches children to speak, read, and write in a second language by surrounding them with conversation and instruction in that language. Note that English immersion may differ from other immersion programs.

**inclusion:** The practice of placing students with disabilities in regular classrooms. Also known as *mainstreaming*.

**independent study:** Specially designed instruction in courses taught through a variety of delivery methods that complement traditional high-school curricula and provide an accredited diploma.

**individual education program (IEP):** A written plan for a student with learning disabilities, created by the student's teachers, parents or guardians, the school administrator, and other interested parties. The plan is tailored to the student's specific needs and abilities, and outlines goals for the student to reach. The IEP should be reviewed at least once a year.

**instructional minutes:** The amount of time the state requires teachers to spend providing instruction in each subject area.

**integrated curriculum:** The practice of using a single theme to teach a variety of subjects. It also refers to an interdisciplinary curriculum, which combines several school subjects into one project.

**international baccalaureate (IB):** A rigorous college preparation course of study that leads to examinations for highly motivated high-school students. Students can earn college credit from many universities if their exam scores are high enough.

**International Standard Classification of Education (ISCED):** Used to compare educational systems in different countries. ISCED is the standard used by many countries to report education statistics to the United Nations Educational, Scientific, and Cultural Organization (UNESCO) and the Organization for Economic Co-operation and Development (OECD). ISCED divides educational systems into seven categories, based on six levels of education.

**intervention:** The funds that schools get for students who are not learning at grade level. They can be used to fund before-school or after-school programs or to pay for materials and instructors.

**job shadowing:** A program that takes students into the workplace to learn about careers through 1-day orientations or more extensive internships to see how the skills learned in school relate to the workplace.

**joint school districts:** School districts with boundaries that cross county lines.

**language arts:** Another term for English curriculum. The focus is on reading, speaking, listening, and writing skills.

**magnet school:** A school that focuses on a particular discipline, such as science, mathematics, arts, or computer science. It is designed to recruit students from other parts of the school district.

**mainstreaming:** The practice of placing students with disabilities in regular classrooms; also known as *inclusion*.

**manipulatives:** Three-dimensional teaching aids and visuals that teachers use to help students with math concepts. Typical tools include counting beads or bars, base-ten blocks, shapes, fraction parts, and rulers.

**master's degree:** A degree awarded for successful completion of a program generally requiring 1 or 2 years of full-time college-level study beyond the bachelor's degree. One type of master's degree, including the Master of Arts degree or MA, and the Master of Science degree or MS, is awarded in the liberal arts and sciences for advanced scholarship in a subject field or discipline and demonstrated ability to perform scholarly research. A second type of master's degree is awarded for the completion of a professionally oriented program, for example, an MEd in education, an MBA in business administration, an MFA in fine arts, an MM in music, an MSW in social work, and an MPA in public administration. Some master's degrees—such as divinity degrees (MDiv or MHL/Rav)—may require more than 2 years of full-time study beyond the baccalaureate degree.

**minimum day:** A shortened school day that allows teachers to meet outside of the classroom and work together.

**modernization:** The installation of new plumbing, air conditioning, floors, cabinets, carpeting, and so on, on school grounds.

**multiple-subject credential:** A credential required to teach in elementary and middle-school classrooms. It qualifies a teacher to teach multiple subjects in a self-contained class.

**national blue ribbon award:** This award honors public and private K–12 schools that are academically superior or that demonstrate dramatic gains in student achievement.

**norm-referenced assessment:** An assessment in which an individual or group's performance is compared with that of a larger group. Usually the larger group is representative of a cross-section of all U.S. students.

**open court reading series:** A program that provides systematic, explicit instruction to help students learn the structure of words and how to sound them out. Fluent reading and comprehension by the end of first grade is a program goal.

**parcel tax:** An assessment on each parcel of property that must be approved by two-thirds of the votes in a school district. The proceeds are generally used for educational programs, not for construction or renovation, which is normally financed through a general obligation bond measure.

**parent teacher association (PTA):** A national organization of parents, teachers, and other interested persons, which has chapters in schools. It relies entirely on voluntary participation and offers assistance to schools in many different areas.

**peer assistance and review program (PAR):** A program that encourages designated consulting teachers to assist other teachers who need help in developing their subject-matter knowledge, teaching strategies, or both. They also help teachers to meet the standards for proficient teaching.

**peer resource program:** A program that trains students to provide their peers with counseling, education, and support on issues such as prejudice, drugs, violence, child abuse, dropping out, and peer pressure. Students are also trained to provide tutoring and conflict mediation.

**percentile ranks:** One way to compare a given child, class, school, or district to a national norm.

**phonics:** An instructional strategy used to teach reading. It helps beginning readers by teaching them letter–sound relationships and having them sound out words.

**physical education (PE):** Activities focused on developing physical and motor fitness; fundamental motor skills and patterns; and skills in aquatics, dance, individual and group games, and sports (including intramural and lifetime sports). The term includes special PE, adaptive PE, movement education, and motor development.

**portable:** A term commonly used to describe single-unit, relocatable buildings. A portable building can be moved from one site when it is no longer needed and used again in another location.

**portfolio:** A collection of various samples of a student's work throughout the school year, which can include writing samples, examples of math problems, and results of science experiments.

**postbaccalaureate enrollment:** The number of students working toward advanced degrees and of students enrolled in graduate-level classes but not enrolled in degree programs. See also *graduate enrollment.*

**postsecondary education:** The provision of formal instructional programs with a curriculum designed primarily for students who have completed the requirements for a high-school diploma or equivalent. This includes programs of an academic, vocational, and continuing professional-education purpose, and excludes vocational and adult basic-education programs.

**primary language:** A student's first language or the language spoken at home.

**private school:** Private elementary/secondary schools surveyed by the Private School Universe Survey (PSS), which are assigned to one of three major categories (Catholic, other religious, or nonsectarian) and, within each major category, one of three subcategories based on the school's religious affiliation provided by respondents.

**professional development:** Programs that allow teachers or administrators to acquire the knowledge and skills they need to perform their jobs successfully.

**professor:** In the context of this book, a person who disseminates knowledge to students at a university (at any level) and a person who has an intimate knowledge of and experience in the subject being taught.

**professor (rank):** A faculty member at a university who has reached the top level teaching rank; typically a tenured position.

**proficiency:** Mastery or ability to do something at grade level.

**program improvement (PI):** A multistep plan to improve the performance of students in schools that did not make adequate yearly progress under the No Child Left Behind initiative for 2 years in a row. Only schools that receive federal Title I funds may be entered in Program Improvement. The steps in

PI can include a revised school plan, professional development, tutoring for some students, transfer to another school with free transportation, and, at the end of 5 years, significant restructuring.

**pull-out programs:** Students receive instruction in small groups outside of the classroom.

**pupil–teacher ratio:** The total student enrollment divided by the number of full-time equivalent teachers. The pupil–teacher ratio is the most common statistic for comparing data across states; it is usually smaller than average class size because some teachers work outside the classroom.

**regional occupational programs (ROP):** State-funded programs for job training, jobs-related counseling, and skills upgrades for students ages 16–18. Students often take ROP classes in high school to start learning a trade.

**regular school:** A public elementary/secondary school providing instruction and education services that does not focus primarily on special education, vocational/technical education, or alternative education, or on any of the particular themes associated with magnet/special-program–emphasis schools.

**resource specialists:** Specially credentialed teachers who work with special-education students by assisting them in regular classes or pulling them out of class for extra help.

**resource teacher:** A teacher who instructs children with various learning differences. Most often these teachers use small group and individual instruction. Children are assigned to resource teachers after undergoing testing and receiving an IEP.

**rubric:** Refers to a grading or scoring system. A rubric is a scoring tool that lists the criteria to be met in a piece of work. A rubric also describes levels of quality for each of the criteria. These levels of performance may be written as different ratings (e.g., Excellent, Good, Needs Improvement) or as numerical scores (e.g., 4, 3, 2, 1).

**safe harbor:** An alternate method for a school to meet AMO if it shows progress in moving students from scoring at the "below proficient" level to the "proficient" level or above on STAR, CAHSEE (the California High School Exit Examination) and/or CAPA (the California Alternate Performance Assessment). The state, school districts, and schools may still make AYP if each subgroup that fails to reach its proficiency performance targets reduces its percentage of students not meeting standards by 10% of the previous year's percentage; plus, the subgroup must meet the attendance-rate or graduation-rate targets. (Dataquest).

**standardized achievement test (SAT):** Also known as the SAT Reasoning Test (formerly called the Scholastic Aptitude Test), this test is widely used as a college entrance examination. Scores can be compared to state and national averages of seniors graduating from any public or private school.

**SAT II:** This was formerly known as the *achievement test* and was renamed the SAT II: Subject Test. It is administered by the College Board and widely used as a college entrance exam. Students may take the test more than once, but only the highest score is reported at the year of graduation.

**school accountability report card (SARC):** An annual disclosure report for parents and the public produced by a school, which presents student achievement, test scores, teacher credentials, dropout rates, class sizes, resources, and more. The SARC is required by state and federal law.

**school improvement program (SIP):** A state-funded program that helps elementary, intermediate, and secondary schools to improve instruction, services, school environment and organization at school sites according to plans developed by School Site Councils (see School Site Council).

**school site council (SSC):** A group of teachers, parents, administrators, and interested community members who work together to develop and monitor a school's improvement plan. It is a legally required decision-making body for any school receiving federal funds (see School Improvement Plan).

**scientifically based research:** Research that involves the application of rigorous, systemic, and objective procedures to obtain reliable and valid knowledge relevant to educational activities and programs.

**secondary school:** A school comprising any span of grades beginning with the next grade following an elementary or middle school (usually Grade 7, 8, or 9) and ending with or below Grade 12. Both junior high schools and senior high schools are included.

**sheltered English:** An instructional approach in which classes are composed entirely of students learning English. Students are taught using methods that make academic instruction in English understandable. In some schools, students may be clustered in a mainstream classroom.

**single-subject credential:** A credential required to teach middle or high school in California. It authorizes a teacher to teach in a single subject area such as English or a foreign language.

**socioeconomically disadvantaged:** Students whose parents do not have a high-school diploma or who participate in the federally funded free/reduced-price meal program because of low family income.

**special day classes:** Full-day classes for students with learning disabilities, speech and/or language impairments, serious emotional disturbances, cognitive delays, and a range of other impairments. Classes are taught by certified special-education teachers. A student may be placed in a regular classroom as appropriate according to the student's IEP.

**special education:** Special instruction provided for students with educational or physical disabilities, tailored to each student's needs and learning style.

**staff development days:** Days set aside in the school calendar for teacher training. School is not generally held on these days.

**standardized test:** A test that is in the same format for all who take it. It often relies on multiple-choice questions, and the testing conditions—including instructions, time limits, and scoring rubrics—are the same for all students, though sometimes accommodations on time limits and instructions are made for disabled students.

**standardized testing and reporting program (STAR Program):** The three tests that are required for grades 2–11.

**standards-referenced tests:** Also known as *standards-based assessments.*

**student:** The recipient of knowledge from the teacher/professor who has the capability of retaining the knowledge and is able to assimilate/sort the knowledge for further thought and practice.

**student study team (also referred to as *student success team*):** A team of educators that comes together at the request of a classroom teacher, parent, or counselor to design in-class intervention techniques to meet the needs of a particular student.

**student teacher:** A teacher in training who is in the last semester of a teacher-education program. Student teachers work with a regular teacher who supervises their practice teaching.

**teacher:** In the context of this book, a person who disseminates knowledge to students at a school (at any level) and a person who has an intimate knowledge of and experience in the subject being taught.

**teaching assistant (school):** A person who assists the teacher or professor in his/her teaching duties—typically a person who has experience in teaching but may no longer be involved in giving classroom instruction on a full-time basis.

**teaching assistant (university):** A person who assists the teacher or professor in his/her teaching duties—typically a graduate research student or graduate research assistant.

**team teaching:** A teaching method in which two or more teachers teach the same subjects or theme. The teachers may alternate teaching the entire group or divide the group into sections or classes that rotate between the teachers.

**tenure:** A system of due process and employment guarantee for teachers. After serving a 2-year probationary period, teachers are assured continued employment in the school district unless carefully defined procedures for dismissal or layoff are successfully followed.

**thematic units:** A unit of study that has lessons focused on a specific theme, sometimes covering all core subject areas. It is often used as an alternative approach to teaching history or social studies chronologically.

**tracking:** The instructional practice of organizing students in groups based on their academic skills. Tracking allows a teacher to provide the same level of instruction to the entire group.

**traditional calendar:** The timetable according to which school starts in September and ends in June for a total of 180 days of instruction.

**traditional public school:** Publicly funded schools other than public charter schools. See also *Public school or institution* and *Charter school.*

**undergraduate students:** Students registered at an institution of higher education who are working in a baccalaureate degree program or other formal program below the baccalaureate, such as an associate's degree, vocational, or technical program.

**university:** An institute of higher education that often consists of a collection of colleges, each of which houses a separate area of scholarship; a community of students and teachers committed to the pursuit of learning, accumulation of knowledge, the transmission of this knowledge to succeeding generations, and the development of new knowledge.

**year-round education:** A modified school calendar that gives students short breaks throughout the year, instead of a traditional 3-month summer break. Year-round calendars vary, sometimes within the same school district. Some schools use the staggered schedule to relieve overcrowding, while others believe the 3-month break allows students to forget much of the material covered in the previous year.

# Index